FUNDAMENTAL EXPERIMENTS in MICROBIOLOGY

KOBY T. CRABTREE, Ph.D.
Professor of Microbiology, University of Wisconsin, Wausau

RONALD D. HINSDILL, Ph.D.
Associate Professor of Bacteriology, University of Wisconsin, Madison

1974 W. B. SAUNDERS COMPANY • Philadelphia • London • Toronto

W. B. Saunders Company: West Washington Square
Philadelphia, PA 19105

12 Dyott Street
London, WC1A 1DB

833 Oxford Street
Toronto, Ontario M8Z 5T9, Canada

Fundamental Experiments in Microbiology ISBN 0-7216-2733-1

Last digit is the print number: 9 8 7 6 5 4 3 2 1

PREFACE

Fundamental Experiments in Microbiology is designed for students taking their first course in microbiology. The main objective is to acquaint the student with some of the principles and techniques which are unique to the field of microbiology. The manual should prove useful for bacteriology majors and for students majoring in natural science, agriculture, sanitary engineering, nursing, and paramedical fields who require a fairly broad introductory course in microbiology. Many of the experiments incorporated in this manual originated in W. B. Sarles et al.: "Experiments in General Bacteriology," 4th ed., 1957—a manual used for many years on the Madison campus of the University of Wisconsin.

Most experiments include a somewhat lengthy introduction to aid the student in understanding the objectives of the experiment and a discussion section at the end to determine his depth of comprehension. This approach, while increasing the size of the manual somewhat, shortens the time required for introductory remarks by the instructor and will be found particularly useful where less experienced teaching assistants must assume much of the teaching responsibility. Although the outline appears to contain a great number of experiments for a semester course, we have found that all of them can be completed if a schedule resembling the one included in the Appendix is followed. Instructors should, of course, feel free to substitute or delete experiments or to treat some or all of the projects in Sections V and VI as optional, to be performed as time and interests dictate.

The first few experiments included in this outline are staining technique, effective use of the microscope in observing bacterial cells, and handling of pure cultures of bacteria, viruses, and fungi. As the manual progresses, physiological and biochemical activities of the bacteria, effects of physical and chemical agents on the growth of microorganisms, and bacterial genetics are presented. The later experiments include aspects of applied microbiology, covering soil, water, and foods, with emphasis on ecology and the sanitary aspects of microbiology.

To gain interest in the applied aspects of microbiology and to learn first hand of problems facing microbiologists and of employment possibilities, students should be encouraged individually or as a group to visit various local industries, public health laboratories, and sewage treatment plants. Students have also reacted very favorably to the last experiment involving the isolation and identification of an unknown, particularly when they have been allowed to bring a sample which is of interest to them. For safety's sake, however, the instructor may wish to exclude potential pathogens by prohibiting samples from dead animals, infected pets, and the like.

KOBY T. CRABTREE
RONALD D. HINSDILL

CONTENTS

Laboratory Suggestions and Rules

SUGGESTIONS

1. It is expected that you will have read and thoroughly acquainted yourself with the subject matter before coming to class. Frequently you will be given a five to ten minute quiz on concepts given in the introduction to the experiment. The examinations may cover objectives, procedures, and results of the experiments and discussion.

2. Laboratory notes should be recorded as you complete the experiment. Accuracy and neatness are considered above artistry. Periodically, your instructor may call for the submission of your results. Be ready to hand them in at any time.

RULES

1. Cleanliness is next to virtue in bacteriology. You are expected to wipe clean your own bench space *before* and *after* the laboratory period with the bacteriocide (Roccal) provided. If you spill stains and reagents on your bench, be sure to remove them immediately.

2. Accidents, such as spilled strong acids and bases, cuts, burns, and swallowed or spilled cultures, should be reported immediately to your instructor.

3. Many of the cultures with which you will be working (including your own isolate) are potentially pathogenic; therefore, you must always practice aseptic techniques in handling and transferring them.

4. Properly label all of the materials to be used in the experiment; unlabeled test tubes and Petri dishes will be discarded by your instructor. All dirty and used glassware should be returned promptly to the receptacle provided. Do not accumulate unnecessary glassware in either your drawer or incubators. All used pipettes should be discarded into the tray or receptacle.

5. Microscopes must always be returned to their proper storage place, generally a locked cabinet, when not in use. Your instructor will examine your

microscope for cleanliness and position of objectives during storage (read Experiment 2 for proper maintenance of microscope). Any damage or defect in the microscope should be reported immediately to your instructor.

6. The common equipment needed for the laboratory will be found in your desk drawer, basket, or equivalent. Be sure to inventory the equipment assigned to you on the first and last day of laboratory. It is your responsibility to replace any lost articles.

SECTION

I

BASIC TECHNIQUES, MICROSCOPY AND CYTOLOGY

EXPERIMENT 1

Staining of Bacteria: The Simple Stains

Unstained bacterial cells are nearly transparent when observed by the conventional light microscope and hence are very difficult to see. Most bacteria, however, readily react with simple stains because the protoplasm of bacteria is basophilic. These simple stains allow one to distinguish bacterial cells of various morphological types (cocci, bacilli, vibrios, spirilla, and so forth) from extraneous material present in the stained smear. In addition, stained cells are firmly fixed on the slide, in contrast to living specimens. Stained slides may be retained as a record for a long period of time. Simple stains such as methylene blue, crystal violet, safranin, and acid and basic fuchsin are very widely used.

MATERIALS

1. Löffler's methylene blue stain.
2. Hucker's crystal violet stain.
3. Ziehl's carbol-fuchsin stain.
4. Pure cultures of *Escherichia coli, Staphylococcus aureus, Bacillus cereus, Spirillum serpens.*

PROCEDURE

1. Preparation of Slides. A clean, grease-free glass slide is essential for obtaining good stained preparations and the slide should be cleaned as follows: Wet the tip of your index finger and rub it on a cake of Bon Ami; spread the paste thus formed over both surfaces of the slide; allow the paste to dry and then wipe it off with a clean paper towel. Cleaned, grease-free slides should be handled with care. Always hold the slide at the edges. Remember, grease from your fingers causes water to collect in tiny droplets and interferes with even spreading of bacterial specimens over the slide. Aseptic techniques for preparing bacterial smears are presented in Figure 1-1.

ASEPTIC TECHNIQUES IN PREPARING BACTERIAL SMEARS

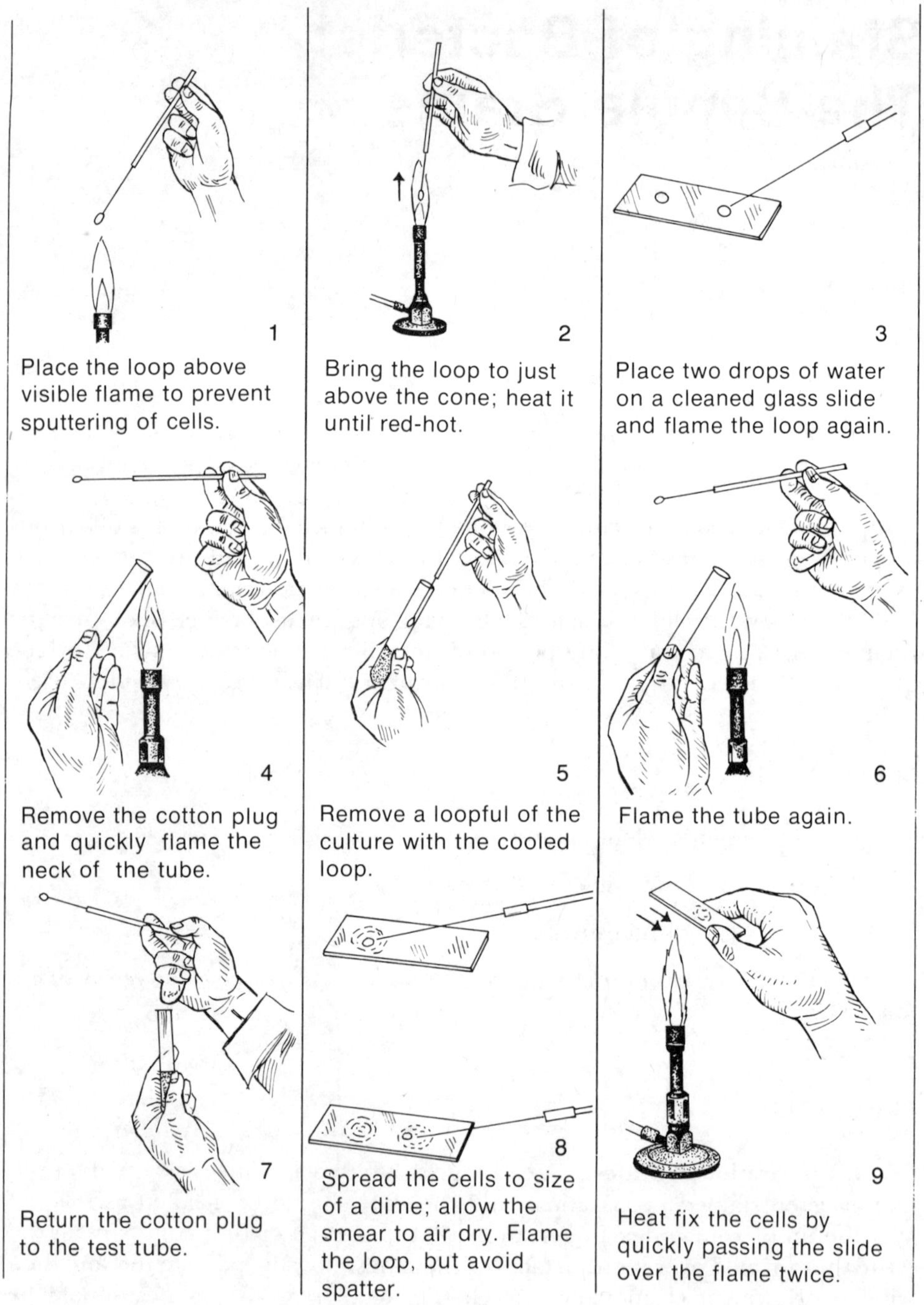

FIGURE 1–1

2. Following Figure 1-1, sterilize the wire inoculating loop by flaming it until it is red hot. Cool for a few seconds. Do not touch it to anything.

3. Using the sterile loop, place two small drops of tap water on the slide approximately 1.5 cm apart.

4. Resterilize the inoculating loop in the flame as shown in Figure 1-1; cool for a few seconds. With the tip of this sterile loop, remove a very small amount of growth from the surface of the *E. coli* agar slant culture. Do not dig into the agar medium. Only a very small amount of the culture is needed to make a smear. (Your instructor will demonstrate the detailed aseptic techniques involved in the handling of pure cultures.)

5. Mix the culture on the tip of the inoculating loop with one of the drops of water on the slide. With a circular motion, spread the drop out to the size of a dime. A well-mixed smear has a *pale,* milky appearance. Flame the loop again as soon as the smear is made by holding the loop above the flame until the inoculum is dry. Only when the inoculum is dry should the loop be plunged into the flame and heated until red hot. If heated correctly, inoculum will not be spattered on your bench top or clothing.

6. Repeat Steps 4 and 5 with the second culture, *S. aureus,* on the other droplet on the same slide. Remember, each time the inoculating loop is used it is essential to flame sterilize it in order to avoid contamination of the succeeding pure culture.

7. On a second grease-free slide, follow the same procedure for cultures of *B. cereus* and *S. serpens.*

8. Allow the smears to air dry, then "heat fix" them by quickly passing the slides through the flame (right side up) two or three times (see Fig. 1-1). Properly "heat-fixed" slides can be handled easily; if they are too hot to handle, you have heated them too much.

9. Place the air-dried, heat-fixed slides on a staining rack over the sink or staining tray; make sure the slides are level.

10. Stain *E. coli* and *S. aureus* smears with crystal violet and *B. cereus* and *S. serpens* smears with Löffler's methylene blue for one to two minutes. Do not allow to dry. (You may consult with a classmate on the selection of the simple stains; your classmate or partner may use carbol-fuchsin in place of crystal violet.)

11. After the smears have been exposed to the stains for 1 to 2 minutes, pour off the excess stain into the sink or staining tray. Wash the slides by placing them under a gentle stream of water or by dipping them in a tumbler of water until the water draining from the slides becomes colorless.

12. Remove all excess water from the slides by touching one corner of each slide to a clean, absorbent paper pad and gently blotting dry, as shown in Figure 1-2. Label the slides with a marking pencil.

The dried slides are now ready for microscopic examination. Keep them in your slide box until you have learned the parts and proper use of the microscope.

SIMPLE STAINING PROCEDURE

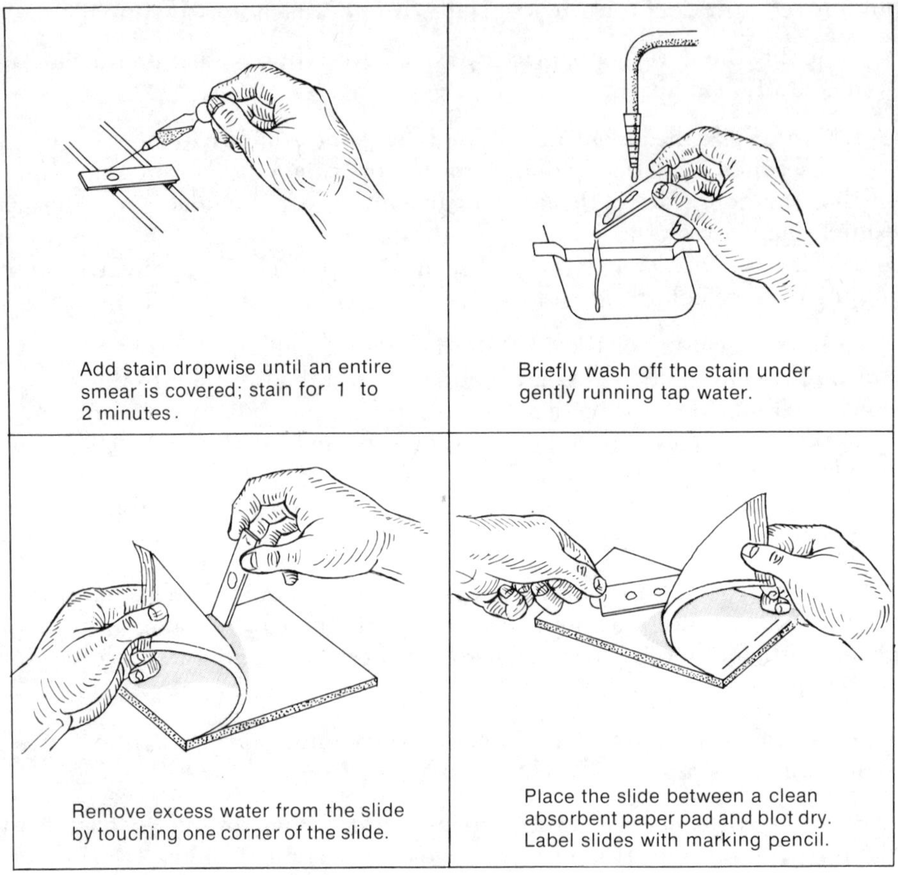

FIGURE 1–2

EXPERIMENT 2

The Construction and Operation of The Microscope

The microscope is one of the most important instruments employed by the microbiologist. To study the morphology of microorganisms such as bacteria, yeasts, molds, algae, and protozoa, you must become thoroughly acquainted with the construction and proper use of the instrument.

The compound microscope used in bacteriology is a precision instrument and its mechanical parts, e.g., calibrated mechanical stage and fine adjustment knob, are easily damaged; all lenses, particularly the oil immersion objective, are delicate and expensive; thus, the instrument must be handled with care. If the microscope must be carried to your bench, use two hands as shown in Figure 2-1 and hold it close to your body so that it cannot be jarred out of your hands even if another person bumps you. Do not tilt the microscope or the ocular lens(es) may fall out.

MATERIALS

1. Compound microscope.
2. Simple stained slides from Experiment 1.
3. Immersion oil.
4. Lens tissue.

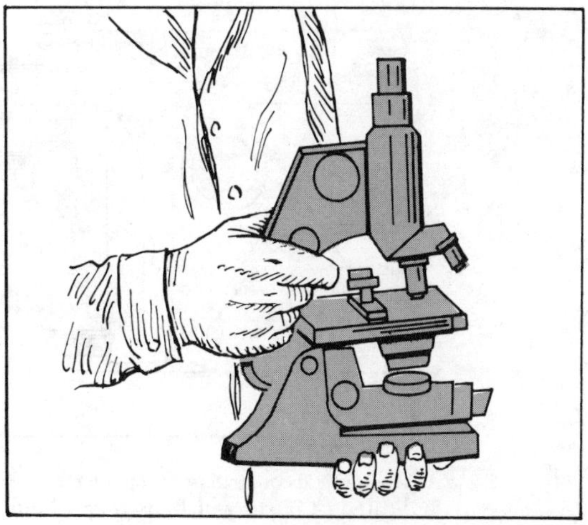

FIGURE 2–1. The proper and safe way to carry a microscope.

PROCEDURE

PART A. EXAMINATION OF THE MICROSCOPE

Examine the microscope to become familiar with its construction. Carefully read the following paragraphs and fill in the names of the various parts in the blanks provided in Figure 2-2.

The Nonoptical Portion of the Microscope. The upper part of the microscope consists of a large movable (or fixed) tube, the *body tube* (1), supported in a vertical position by an erect but curved heavy metal piece called the *arm* (2). The lower part of the arm is the columnar *pillar,* which is attached to the heavy horseshoe-shaped *base* (3). The arm and pillar are connected by a hinge called

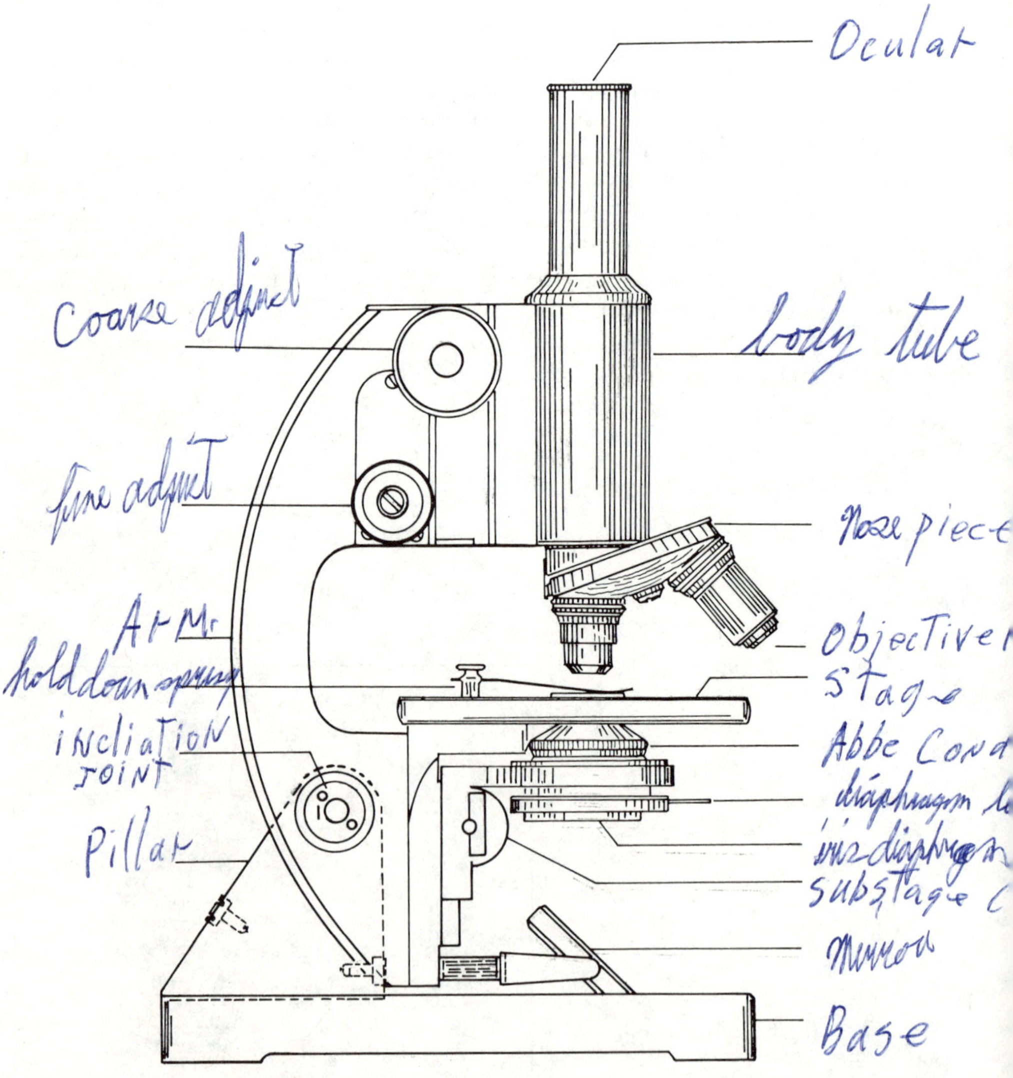

FIGURE 2–2. Mechanical construction of a microscope. Fill in the blanks above with the names of the various parts indicated. (Courtesy of Bausch and Lomb Optical Co.)

the *inclination joint.* (Hold the base with one hand and the arm with the other and gently but firmly pull the arm toward you; the body tube should become inclined toward you. However, you will not be using the microscope in this position and the body tube should be kept vertical to the plane of the base at all times. This will prevent immersion oil from dripping off the slide into the condenser below.) Below the body tube there is a horizontal, solid, fixed *stage* (4). (Some microscopes have a movable stage rather than a movable body tube.) The stage is provided either with hold down springs or, more commonly, with a mechanical stage which facilitates the moving of the glass slide. Mechanical stages (often calibrated) can be moved forward, backward, or sideways over the fixed microscope stage. The mechanical manipulator knobs (one for forward-backward and the other for side movements) are usually located on or immediately below the stage. Locate them and gently move the mechanical stage in all directions by using one knob at a time or two knobs simultaneously.

The Optical Part of the Compound Microscope. The usual source of illumination is a tungsten filament lamp. In many modern light microscopes, substage illumination is by built-in *substage lamp* (either fixed or movable) rather than by an adjustable mirror at the base to reflect light upward (see demonstration table).

Immediately above the built-in substage illuminator or mirror is a series of light control mechanisms. The first one is the *iris diaphragm,* which receives the light from the substage illuminator. Using the diaphragm lever, regulate the amount of light entering the *Abbé condenser.* The function of the condenser (two-glass lens system) is to concentrate the light on a small area just above the stage. Gently rotate the substage condenser knob clockwise and counterclockwise to move the condenser up and down.

Microscopes used in bacteriology are usually equipped with three *objective* lenses which are attached to a revolving *nosepiece* located at the lower end of the body tube.

At the upper end of the body tube there is a lens system known as the *ocular* or eyepiece into which one looks when using the microscope. The function of the ocular is to magnify the image formed by the objectives (see Fig. 2-3). The magnification of the ocular is usually 10×, which means that it magnifies the image formed by the objective ten-fold.

The three objectives are (1) the *low power objective,* which is the shortest of the three and is marked 10×; (2) the *high dry objective,* which is longer and generally has a magnification of 43 to 45×; and (3) the *oil immersion objective,* the longest of the three lenses and generally marked with a distinctive colored or dark band. Magnification is in the range of 95 to 100×. The focal lengths (which may or may not be marked on the lenses) and working distances of a typical set of objectives are shown in Figure 2-4.

The shorter the focal length of an objective lens the less the *working distance;* i.e., the distance between an objective lens and the slide when the specimen is seen in sharp focus. For example, the oil immersion lens (97×) with a focal length of 1.8 mm has a working distance of only 0.14 mm clearance between the lens and the object on the slide. If a cover glass is used, the available working distance will be even less; therefore care must be exercised not to damage the lens.

Focusing of objectives on the object to be viewed can be accomplished by rotating the coarse adjustment and fine adjustment knobs. These two knobs are located on the upper or lower part of the arm and may be separate or coaxial

PATH OF LIGHT THROUGH THE MICROSCOPE

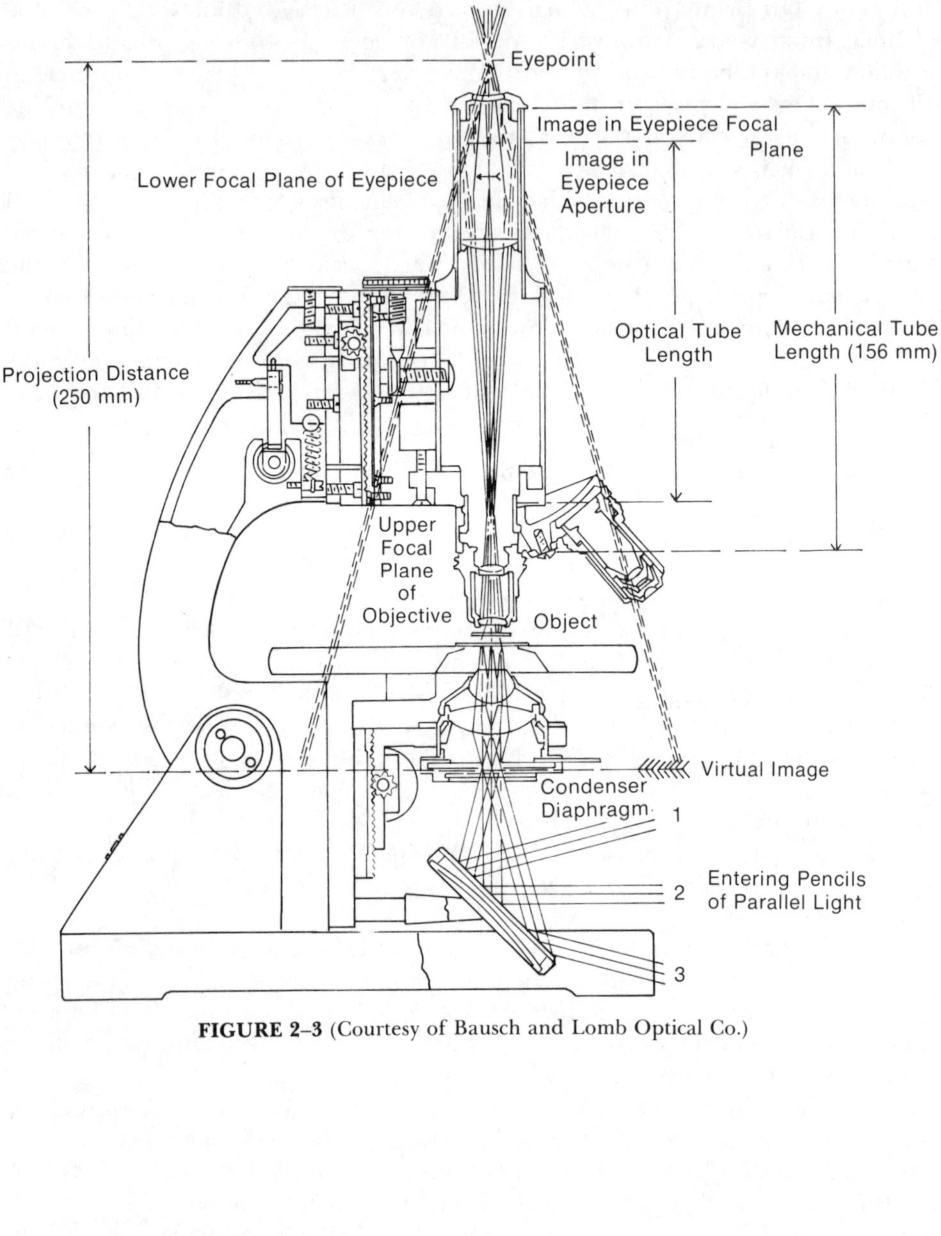

FIGURE 2–3 (Courtesy of Bausch and Lomb Optical Co.)

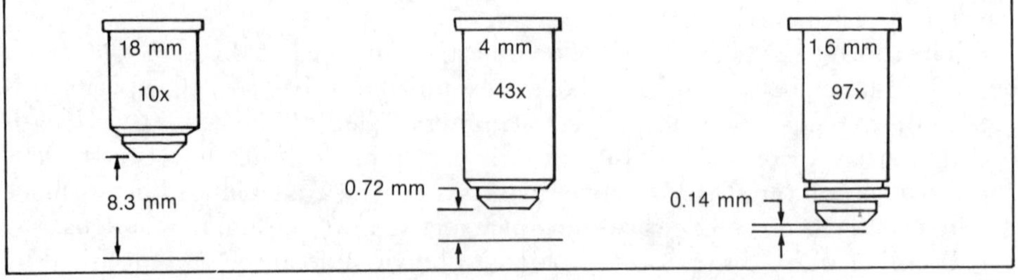

FIGURE 2–4. Focal length and working distance of typical objective lens.

TABLE 2-1.

Objective		Ocular Magnification Number				*Approximate Working Distance mm*
Equivalent Focal Length mm	*Magnification*	*4×*	*5×*	*10×*	*15×*	
16	10×	40	___	___	___	8.3
4	43×	___	___	430	___	0.72
1.8	97×	___	___	___	___	0.14

depending on the brand of microscope. Turn the coarse adjustment knob and see what happens to the body tube with its objectives (or to the stage if your microscope has a movable stage.) Similarly, turn the fine adjustment knob. One revolution of the fine adjustment knob may raise or lower the body tube (or stage) only 0.1 mm; therefore, you will have to watch very closely to detect any change in position of the tube or stage.

After you have become acquainted with the parts, fill in the blank spaces in Figure 2-2 and complete Table 2-1. Remember, the magnifying power of a compound microscope is obtained by multiplying the magnifying power of the objective by the magnifying power of the ocular. After you have completed the figure and table, ask your instructor to test your familiarity with your microscope.

PART B. USE OF THE MICROSCOPE

1. Place the low power objective above the stage by rotating the revolving nosepiece approximately 2.5 cm above the stage. If your microscope has a movable stage, lower the stage approximately 2.5 cm below the tip of the low power objective.

2. Place a slide you have prepared in Experiment 1 on the stage with the stained smear surface facing upward. Press the lever of the mechanical stage located on the left-hand side of the slide. Secure the slide flat against the stage, then release the lever; now the slide is firmly secured by the mechanical stage. By turning the appropriate mechanical stage knobs, move the slide until the stained smear area is located over the center of the light source (circular opening in the stage).

3. Looking at the microscope from the side, lower the objective until the lens is about 0.5 cm above the slide. If your microscope has a movable stage, raise the stage to approximately 0.5 cm below the tip of the low power lens. (Most microscopes have a stop which prevents the low power objective from touching the slide. For this reason, the low power objective is placed above the stage when the microscope is not in use, or during storage.)

4. Looking through the ocular, raise the tube slowly by turning the coarse adjustment knob counterclockwise until the image of the smear comes into approximate focus. With microscopes having a movable stage, slowly lower the stage until the image of the smear comes into approximate focus.

5. Bring the image to sharp focus by turning the fine adjustment knob counterclockwise. For microscopes with a movable body tube, *always focus **up** on an object to avoid any chance of damage to the objective lens. Never attempt to get the image in focus by lowering the body tube.* For microscopes with a movable stage, always focus ***down*** to avoid any chance of damage to the objective lens.

Occasionally you will find dust or lint particles on the lenses of the ocular which interfere with the observation of the object. Such particles, if they are on the ocular lens, will rotate as you rotate the ocular piece. Clean the lens with lens paper provided. *Do not* use any other tissues, such as Kleenex or a handkerchief; they may easily scratch the lens system.

6. After the examination with low power is complete, continue your observation with high power. To move from the sharply focused low power objective to high dry lens, you need only rotate the nosepiece until it is locked in place, since most modern laboratory microscopes are of *parfocal* design. The object under high dry lens will be nearly in focus. The image can be brought to sharp focus by gently rotating the fine adjustment knob (most often counterclockwise, i.e., toward you).

NOTE: Focusing is again achieved by upward movement of the objective (or by lowering of the stage for microscopes with a movable stage). Open and close the diaphragm; what is the effect? Move the condenser up and down to see if better definition can be achieved. In most cases, the condenser should be at its highest point when you are observing stained smears.

7. To move from the sharply focused high dry objective to the oil immersion objective, rotate the high dry lens halfway so that no lens is directly above the smear. At this point place one or two drops of immersion oil directly over the smear and allow to flow or spread over the smear without actually touching the slide with the dropper. Gently rotate the oil immersion objective into the oil. Again, a well-maintained microscope gives you a nearly focused image. If the oil immersion objective ever comes in contact with the slide before it is locked in place, call your instructor for advice. *Never force any parts of the microscope.* Bring the image to sharp focus with the fine adjustment knob as described for the high dry objective. Note the difference in the size and shape of the organisms. Again observe the effect of the diaphragm and the substage condenser.

8. Make a sketch of the four organisms that you have prepared in Experiment 1 in the space provided on page 19.

9. If you have not completed your observation by the end of the class period, save your slides until next period. Before placing the slides in your slide box, soak up most of the excess oil by placing the slide between paper pads. Finish cleaning by dropping a small amount of xylol onto the slide as it is held over the bench sink. Xylol is flammable, so keep your Bunsen burner off while you are cleaning slides.

10. Returning the Microscope to the Cabinet. The microscope should be kept immaculately clean at all times. Check the following list to make sure it is put away properly.

a. Remove the slide from the stage.

b. Rotate the low power objective into position about 2 cm above the stage.

c. Wipe all oil from the oil immersion objective with lens tissue. (Make certain there is no oil on the other objective lenses.)

d. Clean the stage and place the mechanical stage on the center of the stage. Wrap the electrical cord around the base.

When carrying the microscope from the cabinet to your desk, or when returning it, be sure to grip the arm of the instrument with one hand and hold the base with the other hand (see Fig. 2-1). Do not tilt the microscope. Under no circumstances should a student attempt to carry two microscopes at one time.

EXPERIMENT 3

Microscopic Measurement of Bacterial Cells by Use of Ocular And Stage Micrometers

Size and shape are used extensively in identification and classification of bacterial cells. The length and diameter of procaryotic and eucaryotic cells are commonly expressed in micrometers (μm). Formerly the cells were measured in microns (μ), which stood for 1/1000 millimeter, but the unit has been replaced by its equivalent, the μm, which stands for one millionth of a meter (i.e., 10^{-6}).

Most bacteria used in this course will usually measure about 1 to 2 μm in diameter, and up to 3 or 4 μm in length if the bacteria are cylindrical. The major exceptions are a few spirochetes and filamentous bacteria, which may range in length from 5 to as many as 500 μm. However, it should be emphasized that the size of procaryotic organisms in general, or even within a single species, can vary greatly depending on the stage of growth and division. For example, the size of *Mycoplasma* species can range from 0.13 μm diameter to 5 to 15 μm diameter or more.

Therefore, when measuring bacterial cells one should always consider the following factors: (1) kind of medium in which the cells were grown; (2) age of culture (cells preceding division are usually longer than cells immediately after division); (3) whether cells are uniform in shape (coccus, rod, spirochete) or pleomorphic (irregular in shape); (4) whether the cells are unstained (unstained preparations are not subjected to both air drying and heat fixing; as a result they tend to appear larger than heat-fixed, stained cells) or stained, and if stained, with what dye (some dyes stain only the cytoplasm of the cell whereas others stain both the cell wall and cytoplasm, making the cell appear larger). Finally, one should always measure enough cells to insure a random sampling and should express the size as average diameter and length.

Measurements of cells are made by using a scale called an *ocular micrometer* (O) in the eyepiece. The spacing or distance between two lines of the ocular micrometer is arbitrary, as shown in Figure 3-1c. The exact distance between lines of graduations on the ocular micrometer is determined by using a *stage micrometer* (S), a special slide with etched lines that are exactly 0.01 mm, or 10 μm, apart (see Fig. 3-1d).

CALIBRATION OF AN OCULAR MICROMETER

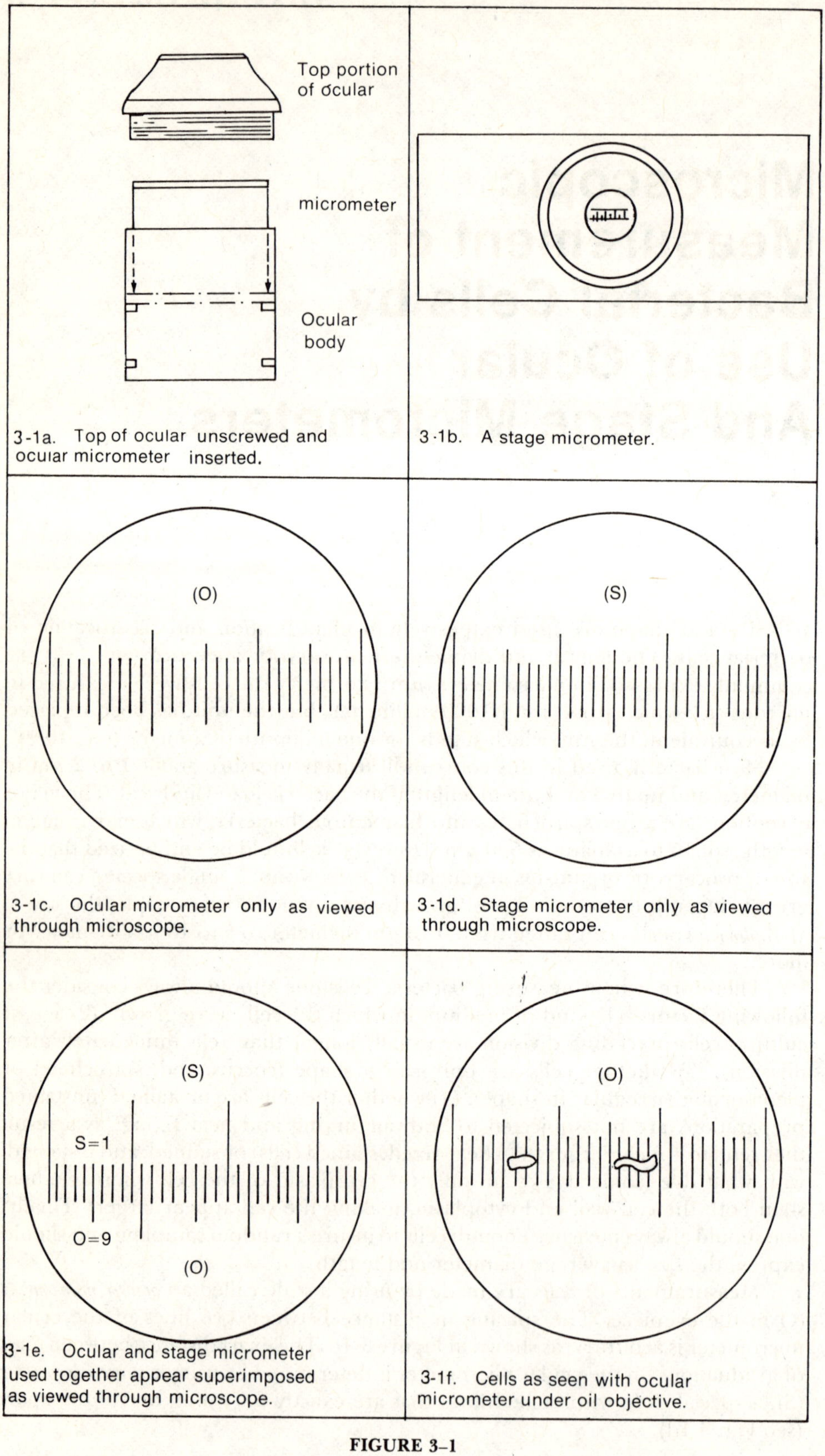

3-1a. Top of ocular unscrewed and ocular micrometer inserted.

3-1b. A stage micrometer.

3-1c. Ocular micrometer only as viewed through microscope.

3-1d. Stage micrometer only as viewed through microscope.

3-1e. Ocular and stage micrometer used together appear superimposed as viewed through microscope.

3-1f. Cells as seen with ocular micrometer under oil objective.

FIGURE 3–1

MATERIALS

Each pair of students will need:

1. Ocular micrometer.
2. Stage micrometer.
3. Prepared slides from Experiment 1.

PROCEDURE

1. Place the ocular micrometer in the eyepiece as shown in Figure 3-1a.

2. Place the stage micrometer on the stage beneath the low power objective (see Fig. 3-1b).

3. Superimpose the ocular micrometer lines over those of the stage micrometer by rotating the eyepiece.

4. After the stage micrometer scale is in clear focus, rotate to the high dry objective. Focus, and then place a drop of oil over the stage micrometer and observe with the oil immersion lens.

5. Superimpose the two scales and determine how many of the ocular graduations are equivalent to one graduation on the stage micrometer. The upper lines (S) represent stage micrometer graduations and the lower lines (0) represent ocular micrometer graduations (Fig. 3-1e).

6. According to the figure, 9 ocular divisions equal 1 stage micrometer division (10 μm); therefore, 1 ocular division equals 10 μm/9, or 1.1 μm.

7. On the basis of the above calibration calculation (i.e., 1 ocular division = 1.1 μm):

   How long is the rod-shaped bacterium in Figure 3-1f?
   How long is the spiral-shaped cell in Figure 3-1f?

8. Place your own slides prepared in Experiment 1 (simple stained slides) as you placed the stage micrometer and measure as many cells as you can. Average the values and record the results in the table on page 25.

RESULTS

AVERAGE SIZE OF CELLS

Simple Stained Cells	*Diameter* μm	*Length* μm
E. coli		
S. aureus		
B. cereus		
S. serpens		

DISCUSSION

1. If 11 ocular divisions are equal to 1 stage micrometer division, then 1 ocular division corresponds to how many μm?

2. Is it better to measure negatively stained cells or positively stained cells? Defend your answer. (You may refer to Experiment 4 which includes the negative stain.)

EXPERIMENT 4

Staining of Bacteria: Differential and Negative Stains

PART A. THE GRAM STAIN

The Gram stain was devised by the Danish scholar, Christian Gram, in 1884. The stain is probably the most widely used one in bacteriology and is valuable because it often enables us to differentiate between two bacterial cultures which are morphologically indistinguishable yet of different species. The stain divides bacteria into two large groups: (1) those that retain the crystal violet dye-iodine complex (violet) throughout the staining procedure, termed gram-positive; and (2) those that lose the crystal violet color complex upon washing with Gram's alcohol but are stained the color of the counterstain (safranin), termed gram-negative. Thus, at the completion of the staining procedure, gram-positive bacteria are stained violet and gram-negative bacteria are pink in color.

The history of the cultures is very important in Gram staining. The age of the cultures and the acidity or alkalinity (pH) of the media in which the bacteria are grown will markedly affect their reaction to Gram's stain. Older cultures of gram-positive organisms (particularly if they show autolysis) and those in acid media frequently appear either gram-negative or gram-variable (both gram-positive and gram-negative cells of the same organisms seen in the same culture). However, gram-negative organisms *do not* become gram-positive regardless of age or medium used. In brief, only a fresh culture should be used for the Gram stain and the student should not be unduly concerned over the presence of a few gram-negative representatives in a predominantly gram-positive culture.

MATERIALS

1. Actively growing cultures (12 to 18 hr) of *Escherichia coli, Staphylococcus aureus, Bacillus cereus,* and *Spirillum serpens.*
2. Gram's crystal violet (see Appendix).
3. Gram's iodine (see Appendix).
4. Gram's alcohol (see Appendix).
5. Safranin (see Appendix).

PROCEDURE

1. Prepare smears (as in Experiment 1) of *E. coli* and *S. aureus* on the same slide. Make sure that the two culture smears are separated by at least 2 cm. Do the same for *B. cereus* and *S. serpens.*

2. Air dry and heat fix in the flame.

3. Place slides on the rack and keep them level. Flood the smears with crystal violet stain for 1 to 2 minutes. Do not allow to dry.

4. Pour off excess stain into the sink and gently wash the slides with tap water.

5. Flood the smears with Gram's iodine solution (acts as mordant, i.e., it is believed iodine reacts with the crystal violet and cell nucleic acid (RNA) to form a complex which is not readily removed from gram-positive cells during the decolorization step). Allow to react for 1 minute. Pour off and treat with iodine again for 1 minute.

6. Wash off the iodine solution with tap water and gently blot the slides dry between clean, absorbent paper pads.

7. Hold the slide at an angle and apply Gram's alcohol, drop by drop, until the stain (violet color) no longer appears in washes from the smears. This generally takes between 10 and 30 seconds. It is important not to overdo the decolorization.

8. Quickly rinse off the alcohol with tap water and blot dry as above.

9. Flood with safranin (counterstain) for 30 seconds.

10. Wash gently with water, drain excess water, and allow to air dry. Mark the slides with wax pencil.

11. Examine both slides under the oil immersion objective without a cover glass. (Remember to use low power objective to locate specimens, high dry to focus, then place a drop of immersion oil, rotate to oil immersion objective and sharpen focus. For microscopes with a movable body tube, always focus ***up*** with the fine adjustment knob to avoid damage to the oil immersion objective. For microscopes with a movable stage, always focus ***down.***)

12. Draw representative fields in the spaces provided.

Possible Mechanism of the Gram Stain. Although the precise mechanism of the Gram stain is not yet known, there are several demonstrable factors associated with the Gram reactions: (1) gram-positive bacterial cell walls may have more affinity for the crystal violet dye-iodine complex than those of gram-negative cells; (2) old, involuted (degenerated) cells or mechanically injured cells or protoplasts (cells with cell wall removed) of gram-positive cells will not retain the dye-iodine complex; (3) extraction of certain lipids and carbohydrates from gram-positive cells makes the cell gram-negative. Thus it appears that the cell wall determines the Gram reaction to a large extent. The gram-positive cells with their thicker cell walls (perhaps also dehydrated by the 95 per cent alcohol) undoubtedly inhibit the passage of the dye complex out of the cell during the decolorization step to a much greater extent than do gram-negative cells: there is a marked difference in destaining rate.

There are considerable differences in the chemical composition and physical structure of the cell walls of gram-positive and gram-negative organisms and these and other physiologically distinguishing characteristics, such as susceptibility to certain drugs and inhibitors, are presented in your text.

PART B. THE ACID-FAST STAIN (ZIEHL-NEELSEN METHOD)

Some bacteria are identified and defined on the basis of a characteristic staining that depends on their lipid-rich cell walls. The cell walls of these organisms, particularly, in the genus *Mycobacterium,* are well endowed with fatty or waxy substances (mycolic acids); thus they do not stain readily under normal staining procedures. Once stained with the aid of heat, they retain dyes with tenacity, that is, they resist decolorization with dilute acid or acid alcohol. Because of this property they are designated "acid-fast" organisms. The mechanism of acid fastness is discussed further in your text.

The object of this experiment is to acquaint you with acid-fast staining and to demonstrate the gross appearance of some organisms grown on an artificial medium.

MATERIALS

1. Pure cultures of *Mycobacterium smegmatis, Bacillus cereus, Escherichia coli.* Cultures of *M. tuberculosis* strain BCG and *M. avium* may be used for demonstration and preparation of phenolized cells.

2. Phenolized sputum sample from a patient, when available (the cells are killed but retain acid fastness).

3. Carbol-fuchsin, acid alcohol, methylene blue.

PROCEDURE

1. Observe the characteristic waxy growth of the mycobacteria on the agar slants and describe the appearance in the space provided.

2. Prepare two slides:

   a. For one slide, use pure cultures of the "nonpathogenic" *Mycobacterium smegmatis* and *E. coli.* Put a small drop of water on one side of a clean slide. Remove a *small amount* of *M. smegmatis* culture with a sterile wire loop and mix thoroughly with the water. Homogenize the lump of bacteria by rotating the wire loop and evenly spread the suspension to the size of a nickel. Flame the wire loop. Be sure to insert the tip of the loop below the cone of the flame (cooler area of the flame) and gradually move into the hotter area. Why? Place a smear of pure culture of *E. coli* (24 hr) on the same slide adjacent to but not touching the first smear. Air dry and heat fix the smears.
   b. For the second slide, use a sputum sample (if available) or

phenolized pure culture of *Mycobacterium* sp. in place of the pure culture of *M. smegmatis;* use *B. cereus* (12 hr) in place of *E. coli.* (*E. coli* represents gram-negative and *B. cereus* represents gram-positive nonacid-fast organisms.) Air dry and heat fix.

3. Place the two slides on a ring stand and place paper towels or a container of some sort under them to protect the bench surface from stain. Flood the slides with carbol-fuchsin and gently heat them. Maintain the heat *just enough* to cause the stain to steam for 4 minutes. Do not allow the stain to dry on the slide during the heating. Keep the slides covered with the stain by adding more fresh stain as water evaporates.

4. Gently wash the slides with water and drain off the water by touching the corner of the slide with an absorbent towel.

5. Decolorize the slide with acid alcohol (3 ml hydrochloric acid in 97 ml of 95 per cent ethyl alcohol) until the wash draining from the slide is colorless, usually 20 to 30 seconds.

6. Wash with water; drain dry.

7. Counterstain with methylene blue for 1 minute.

8. Wash with water and gently blot dry.

9. Examine the slides with oil immersion objective. Sketch the cells from representative fields in the space provided.

PART C. THE NEGATIVE STAIN

As the name implies, this is not a staining of bacteria, but a method of staining the background a solid black. Under the oil immersion objective, the organisms appear unstained on a dark background. Therefore, negative stains are useful in studying morphology and in observing organisms such as spirochetes that cannot be satisfactorily stained by ordinary simple stains. However, it should be cautioned that cells prepared by this method may be subject to either shrinkage or distortion during drying of the stains.

MATERIALS

1. Dorner's nigrosin solution (see Appendix).
2. Pure cultures of *Bacillus cereus* and *Spirillum serpens.*

PROCEDURE

1. Refer to Figure 4-1 and transfer a loopful of the *B. cereus* suspension to one end of a grease-free slide. (Do not allow to dry.)

2. Place one drop of nigrosin solution adjacent to the bacterial suspension.

3. Hold the slide with the drops in your left hand, take one end of a

METHOD FOR SPREADING NIGROSIN-CELL SUSPENSION

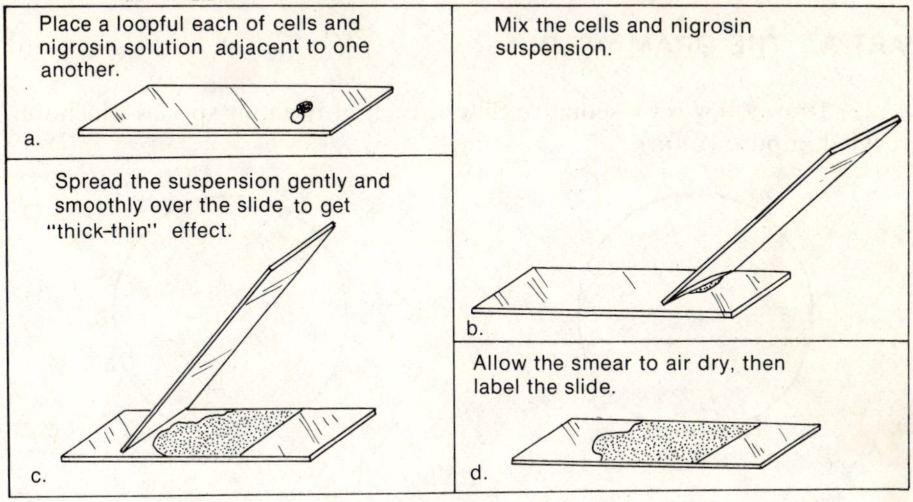

FIGURE 4–1

clean slide in the other hand, and mix the drops together as shown. Gently push the spreader slide over the surface to insure a gradual thinning of the preparation.

4. Allow the slide to dry. Wash the spreader slide.

5. Follow Steps 1 through 4 for *S. serpens.*

6. Examine the slides under the oil immersion objective as for the Gram stain.

7. Draw representative fields in the spaces provided.

8. If you have time, use the ocular micrometer and measure the length of the cells as in Experiment 3.

RESULTS

PART A. THE GRAM STAIN

1. Draw a few representative cells of each of the four species which were provided in pure culture.

−

Species name: Escherichia Coli

+

Species name: Staphylococ aureus

+

Species name: Bacillus subtilis

−

Species name: Rhodospirill rubrum

PART B. THE ACID-FAST STAIN (ZIEHL-NEELSEN METHOD)

1. Describe growth characteristics of mycobacteria on agar slants.

2. Sketch acid-fast stained cells from representative fields.

purple

M. smegmatis

E. coli

Phenolized sputum sample

B. cereus

PART C. THE NEGATIVE STAIN

1. Draw a representative field of each of the two species provided.

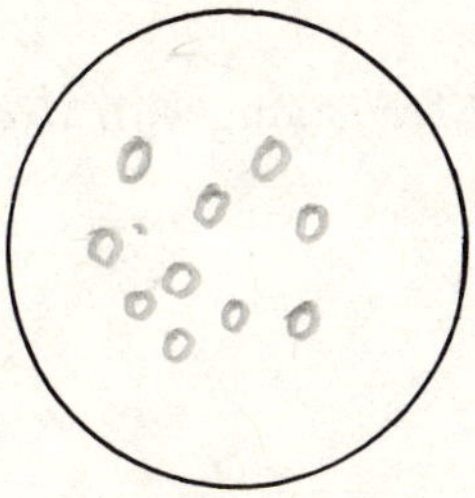

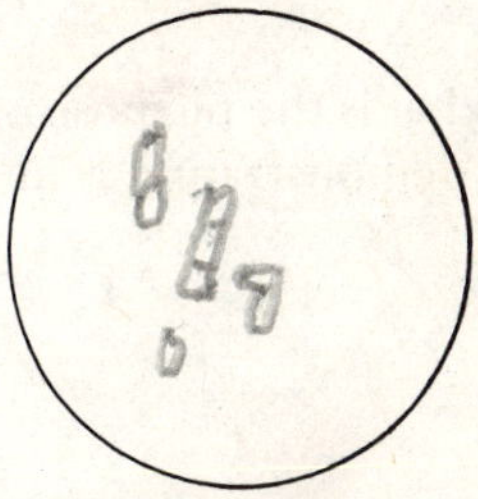

Species name: Strep Lactic

Species name: B. Subtilis

UNK

some look rod pairs of cocci

DISCUSSION

1. What is the purpose of Gram's alcohol in the Gram staining procedure?

The alcohol is used for decoloration of the bacteria (this occurs almost complet with the G^- and to a far lesser extent with

2. What would happen if the Gram iodine step was omitted from the total Gram staining procedure?

The Gram's crystal violet would wash out of both the Gram's Positive and Gram's negative bacteria.

3. What is the purpose of using *B. cereus* or *E. coli* along with *Mycobacterium* species in demonstrating acid fastness?

4. What is the reason for using methylene blue rather than safranin as a counterstain in the Ziehl-Neelsen procedure?

5. What is the advantage of negative stain over all other stains in measurement of bacterial cells?

EXPERIMENT 5

Staining of Bacteria: Cytological Stains

Although a great deal of knowledge concerning the internal structures of procaryotic cells has been obtained through electron microscopy, students still find it interesting (and fun!) to see how much information can be gained through the use of classic staining procedures and the light microscope. Staining methods have been developed for the principal structural elements of bacteria (capsule, cell wall, nuclear area, spore) as well as for various materials found within cells (fats, carbohydrates, volutin, i.e., polymetaphosphate). Unfortunately, these stains work well with only certain types of bacteria, and many of the staining procedures require experience and patience for acceptable results.

For this experiment we have selected three cytological stains and a capsule stain which students will find easy to use.

PART A. BACTERIAL SPORE STAIN

Certain species of rod-shaped bacteria belonging principally to genera *Bacillus* and *Clostridium* develop endospores (so-called because the structure is formed within the cell). As the cell ages and disintegrates, the endospore is released and becomes a mature or free spore. Each vegetative cell produces only one spore; therefore endospore formation (sporulation) is not a multiplication process. Sporulation usually occurs as the culture becomes older, i.e., enters its late logarithmic growth phase.

Spores are extremely resistant to physical agents such as freezing, heating, and desiccation as well as to various chemical agents such as antibiotics or bactericides. As a rule, free spores resist ordinary dyes such as methylene blue, crystal violet, and carbol-fuchsin. However spores, once stained by a specific dye, will resist decolorization by various solvents. This is not true for endospores at early stages of development or for germinating free spores, since they stain like vegetative cells.

MATERIALS

1. Vegetative (12 to 14 hr) and sporulated (24 to 30 hr) cultures of *Bacillus cereus* and *B. polymyxa* on nutrient agar slants.

2. 5 per cent aqueous malachite green solution.
3. Gram safranin stain.
4. Ring and ring stand.
5. Tumblers.

PROCEDURE

1. Select a clean slide and prepare a smear of the young vegetative culture on the left and a smear of the sporulated culture on the right. Work in pairs. One student will use *B. cereus;* the other will use *B. polymyxa.*

2. Air dry and heat fix.

3. Using a wax pencil, trace on all four sides of the slide to build a dike all around the slide (see Fig. 5-1). This procedure is to aid in the retention of the spore stain in relatively large amounts, without spilling over the sides.

4. Place the ringholder on your bench and attach a ring approximately 25 cm from the bench surface. Place a tray or double thickness of paper towels immediately under the ring. This is to protect the bench surface in the event the stain is spilled.

5. Place the slide on the ring; the slide should be tangent to the rim. Ask your instructor if you have difficulty. Make sure that the slide is level.

PROCEDURE FOR SPORE STAINING

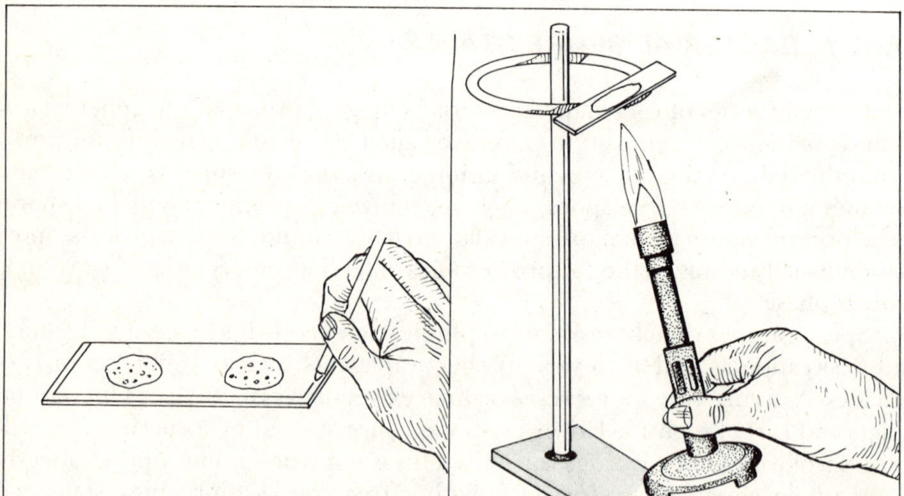

Build a wax pencil dike all around the slide to retain the malachite green.

Gently heat the slide until a trace of steam appears over the stain. Do not boil.

FIGURE 5–1.

6. By drops, add aqueous malachite green over the surface of the slide until the entire slide is flooded.

7. Gently heat the underside of the slide with a Bunsen burner until a trace of steam appears over the stain (do not boil) and maintain this condition. Add stain and continue to heat for a total of about 10 minutes. Do not allow to dry.

8. After 10 minutes of staining, remove the slide with forceps (as this stain is very difficult to clean from your fingers) and wash in a tumbler of tap water for 30 seconds or so. Avoid overwashing. Blot dry.

9. Counterstain for 30 to 45 seconds with Gram's safranin, then wash. Do not overwash as water is a decolorizing agent. Blot dry and examine under oil immersion objective and record results in the space provided.

10. Also examine your partner's slide and note any differences.

A satisfactory preparation will show the spores stained green and the remainder of the cell light pink. Note the location of the spores within the cells. If the spore is located in the center of the cell, it is called spore *central;* if it is slightly off center, it is called spore *subterminal,* and if it is at the end of the cell, it is called spore *terminal.* The size of the endospore and its position in the cell often aid in distinguishing spore-forming species.

PART B. STAINING OF RESERVE MATERIALS IN BACTERIAL CELLS

Many microorganisms store reserve food materials as granules in cells. The stored material can be classified in three large chemical groups: (1) polysaccharides, (2) lipids, and (3) inorganic elements and compounds. Polysaccharide reserve materials are glycogen, glycogen-like, or starch-like materials, and are often accumulated in members of coliform bacteria, i.e., *Escherichia coli* and *Enterobacter (Aerobacter) aerogenes.* These materials are generally considered as iodophilic and stain deep reddish-brown with Lugol's iodine. Some bacteria accumulate chloroform-soluble polyesters of β-hydroxybutyric acid (poly-β-hydroxybutyric acid) which have very high affinity for fat-soluble dyes such as Sudan black-B; hence, the polyester granules are often called sudanophilic granules. However, accumulation of true fats has been observed in many yeast cells. Other bacteria accumulate large reserves of inorganic material believed to consist primarily of metaphosphate granules known as *volutin.* The volutin granules often stain a reddish color with old methylene blue; thus they have also been called *metachromatic granules.* Purple sulfur bacteria, such as *Chromatium,* accumulate elemental sulfur in granular form. Both organic and inorganic granules serve as reserves of carbon and energy; some granules, such as volutin, function as phosphate reserves.

These stored materials can be readily demonstrated under proper conditions and with appropriate staining procedures. You will perform only one of these demonstrations, namely of sudanophilic granules.

MATERIALS

1. Pure cultures of *Spirillum serpens, Torulopsis utilis* (yeast).
2. Sudan black-B staining solution (see Appendix).
3. Xylol.
4. Gram's safranin.

PROCEDURE

1. Prepare smears of *T. utilis* and *S. serpens* on the same slide, one culture smear to the left of center and the other to the right of center (make sure the two smears are well separated).

2. Air dry and heat fix.

3. Flood smears with Sudan black-B solution and allow to react for 5 minutes. Do not allow to dry.

4. Pour off excess stain into staining tray or bench sink and blot dry.

5. Wash the slide with xylol and blot dry.

6. Counterstain with Gram's safranin for 15 seconds and wash with tap water. Blot dry and examine under oil immersion.

If properly stained, the cytoplasm will be stained pink and the granules will appear as bluish-black globules in the cell. Record the results in the space provided.

PART C. STAIN FOR BACTERIAL CELL WALL

This staining procedure works best with large gram-positive cells. Good results will be obtained by most students if directions are followed carefully. Read through the directions before starting so that you can move quickly.

MATERIALS

1. Overnight slant culture of *Bacillus cereus.*
2. Phosphomolybdic acid solution (1.0 per cent).
3. Aqueous solution of methyl green (1.0 per cent).

PROCEDURE

1. Emulsify the growth on the slant with the small amount of condensate at the bottom of the slant (add several drops of water if slant is dry). The object is to prepare a very heavy suspension of cells.

2. Transfer 5 loopfuls of this heavy suspension to one end of a clean slide. Take a second clean slide and, using the edge, make a thick-thin smear as shown in Figure 4-1.

3. Immediately, before the smear can dry, place the slide on the staining rack and flood with phosphomolybdic acid.

4. After 4 minutes, use forceps to tilt the slide and allow the acid solution to drain into the sink or staining tray.

5. Flood the slide with methyl green stain and allow to stand for 5 minutes.

6. Pour off stain and decolorize by gently washing with water. Try not to remove smear, but continue washing until only negligible amounts of methyl green stain appear in the wash water.

7. Allow the slide to air dry. Do not blot.

8. When thoroughly dry, examine under the oil immersion objective. Examine various areas of the smear, starting at the thick end, until a properly stained cell is found. The cytoplasm should be pale green or almost clear, while the cell wall should be dark greenish-blue.

9. Record your results in the space provided.

PART D. DEMONSTRATION OF BACTERIAL CAPSULES

Some bacteria and blue-green algae secrete pronounced slimy or gummy materials on their cell surfaces. If such materials surrounding the cells are compact and appear as a definite shape (circular or oval), they are called a *capsule,* but if they are irregular in shape and loosely attached to the cell, they are called a *slime layer.*

Capsules and slimes are not essential to the life of the cells, but may serve as reserve foods, protection against ingestion by phagocytes, or protection against dehydration. The ability to produce capsules is a genetic trait, but production is profoundly affected by the kind of medium in which the cells are grown. In some bacteria the presence of capsules is indicative of virulence (*Streptococcus pneumoniae* or pathogenic pneumococci), and diagnosis of pneumonia and other diseases is facilitated by determining the presence of capsules. All bacterial capsules appear to be water-soluble.

The chemical composition of capsules varies according to the organism. Some bacteria may produce polymers of glucose (e.g., dextran by *Leuconostoc mesenteroides*), polymers of amino sugars (e.g., hyaluronic acid by pyogenic staphylococci), polypeptides (e.g, polymers of D-glutamic acid by *Bacillus anthracis*), and some are polysaccharide-protein-complex (e.g. dysentery bacilli).

Unstained bacterial capsules are very difficult to observe with the ordinary optical microscope because they are colorless and have low refractive indices. Most capsules have little affinity for the dyes commonly used to stain bacteria; hence special stains are used to demonstrate them. The object of this experiment is to demonstrate the capsules on certain bacteria by positive and negative staining methods.

MATERIALS

1. 48 hr cultures of *Enterobacter aerogenes* and *Staphylococcus epidermidis* on TS agar slants.
2. Alcian blue capsule stain (see Appendix for formula).
3. Carbol-fuchsin stain (Ziehl).
4. India ink (filtered).

PROCEDURE

Demonstration by Positive Staining

1. With a sterile loop, place two small drops of tap water on a slide approximately 2 cm apart. Prepare a very thin smear of *E. aerogenes* to the left of center and *S. epidermidis* to the right of the center. *Note:* to avoid transferring an excess number of cells, use a flame-sterilized needle rather than a loop. Very gently mix the cells with the water on the slide. Remember, bacterial capsules may be water-soluble.

2. Air dry but *do not* heat fix.

3. Flood the smears with alcian blue capsule stain for 1 minute. Wash the slide gently and quickly drain and blot dry.

4. Counterstain with carbol-fuchsin for 30 seconds. Wash quickly and air dry. Examine the smears under the oil immersion objective. Record your observations in the space provided. Note: *S. epidermidis* does not possess capsules but cells of *E. aerogenes* (stained reddish-blue) are surrounded by blue stained capsules.

Demonstration by Negative Staining

1. Place a loopful of water on a slide and transfer a very small number of *E. aerogenes* cells with a sterile inoculating needle. Gently emulsify the cells.

2. Add a loopful of India ink with a sterile loop, then gently mix the ink and cell suspension.

3. Place a cover glass over the mixture and examine the specimen under oil immersion objective. The cells and capsules are unstained because the blackish colloidal particles will not penetrate the interior of the cells or the capsules. Reduce the amount of light and bring a bacterial cell (not capsule) into critical focus. Unstained halo-like material surrounding the cells represents the capsule. Draw a typical field showing the encapsulated cell in the space provided.

4. Repeat the process with *S. epidermidis.* Note colloidal particles of India ink bouncing against the cell surface. Why? What is the name applied to the motion of the India ink particles? Draw a representative field in the space provided.

RESULTS

PART A. BACTERIAL SPORE STAIN

1. Draw a few representative vegetative cells, cells with an endospore, and free spores.

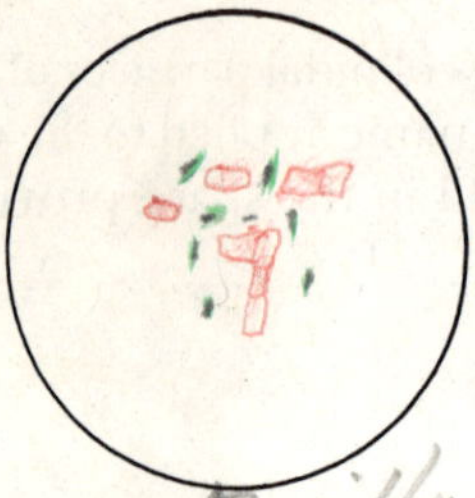

Age of culture: ____________________
Species name: ____________________

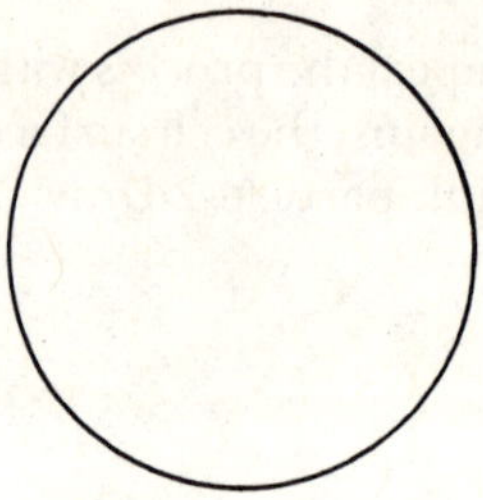

Age of culture: ____________________
Species name: ____________________

PART B. STAINING OF RESERVE MATERIALS IN BACTERIAL CELLS

1. Draw a few representative cells with sudanophilic granules.

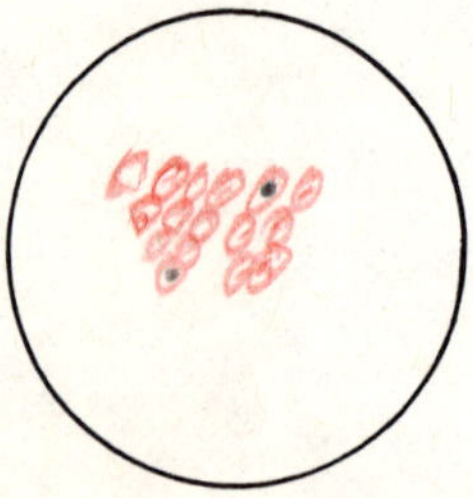

Species name:

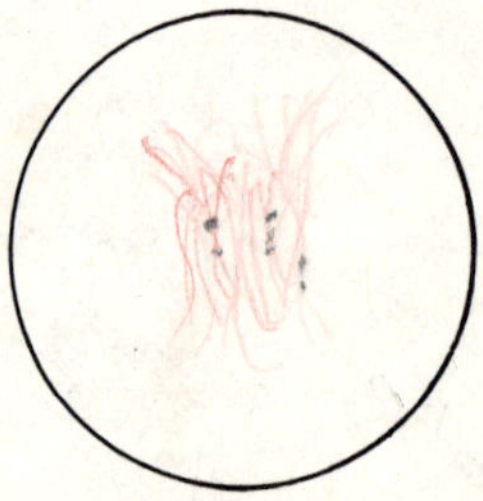

Species name:

PART C. STAIN FOR BACTERIAL CELL WALL

1. Record your observations on the cell wall stain.

PART D. DEMONSTRATION OF BACTERIAL CAPSULES

1. Sketch a representative field of alcian blue stained smears.

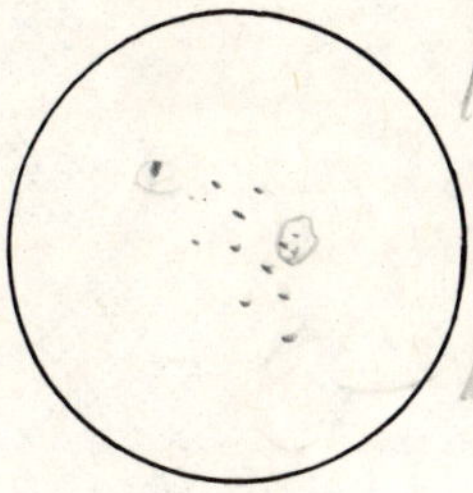

E. aerogenes

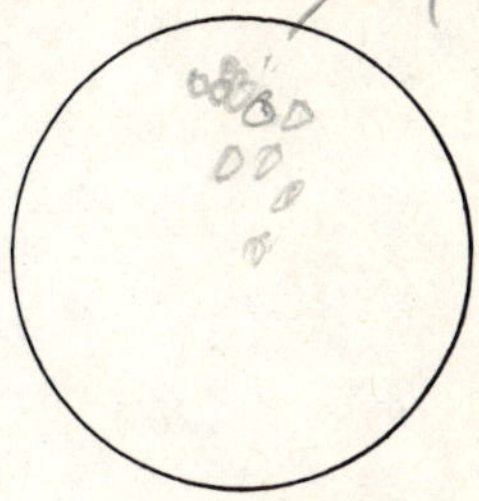

S. epidermidis

2. Sketch a representative field of India ink stained smears.

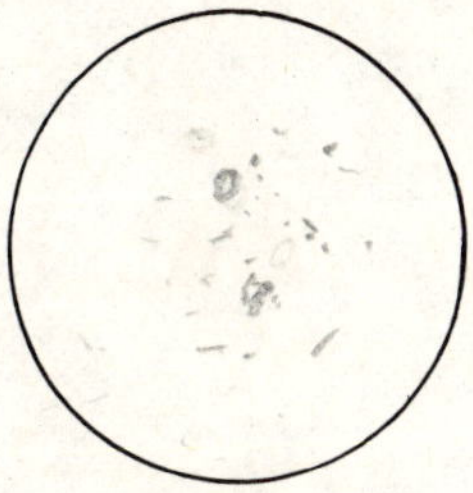

E. aerogenes

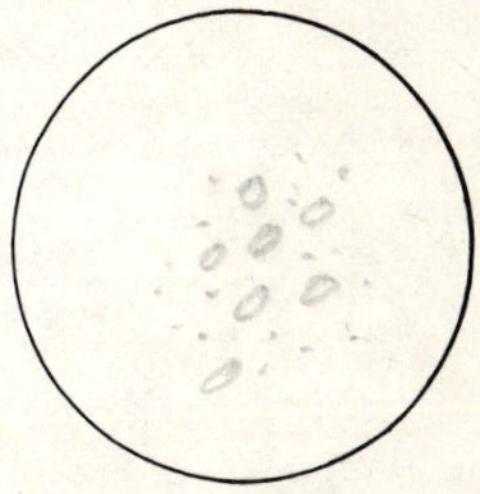

S. epidermidis

DISCUSSION

1. Is spore formation among procaryotic cells a means of multiplication? Explain your answer. What about spore formation among eucaryotic cells?

2. For what purposes do bacteria accumulate intracellular inclusions, e.g., volutin?

3. Must you perform a *spore* stain to tell whether or not a culture is a spore former?

4. How would you demonstrate that capsules may serve as reserve food?

5. A certain bacterial capsule may be used as a blood substitute. What is the name of the product and what kind of microorganism may be able to synthesize it?

Dextran polyglucose

EXPERIMENT 6

Motility of Microorganisms

Some microorganisms are capable of true motility and thus can move under their own power; others are nonmotile. Flagella are the organs of locomotion of bacteria, certain algae, and certain protozoans. Other protozoans may move by means of cilia or pseudopodia (ameboid movement).

Another type of motion characteristic of all small particles suspended in liquid or air is "Brownian movement" or false motility. This is a random, vibratory motion of the particles resulting from their continuous bombardment by invisible, small molecules present in the liquid or air. Students must learn to distinguish true motility from Brownian movement.

Flagellation is a stable inheritable trait in bacterial cells. Along with other morphological and biochemical properties, it is looked upon as an important characteristic for purposes of identification and classification. The flagella may be present singly or in tufts, at one or both ends of the cells, or distributed randomly over the surface of the cell (peritrichous). Further details on the terminology, fine structure, and antigenicity of flagella will be found in the text.

Bacterial flagella are of variable length up to 70 μm, but are only 0.01 to 0.03 μm in width. Thus, while they may be readily observed with the electron microscope, they are too small to be seen with the light microscope. Special stains are available containing a mordant (such as tannic acid) which binds to the flagella and increases their diameter so that they can be viewed in stained smears. These stains are difficult to make, however, and for many purposes microbiologists need only determine whether the culture is motile (an indirect indication of flagellation, but not a clue to the arrangement of flagella on the cell).

Bacterial motility, or the lack of it, is generally ascertained by microscopic examination of the culture in hanging drop or wet mount, by the use of motility agar, or by flagella stain. In this experiment you will compare the first three methods using both known and unknown cultures.

PART A. MOTILITY DEMONSTRATED BY HANGING DROP SLIDE

MATERIALS

1. Hay infusion (cut grass or hay placed in water and allowed to stand at room temperature for 2 to 5 days).

2. Pure cultures (12 to 14 hr) of *Proteus vulgaris, Staphylococcus aureus,* and *Bacillus cereus.*

3. One hanging drop depression slide; cover glass.

4. Vial of Vaseline.

PROCEDURE

1. Clean the depression slide and cover glass by blowing on them and wiping them clean with a piece of cheesecloth.

2. Spread a thin film of Vaseline on one edge of the palm of your left hand.

3. As shown in Figure 6-1a, hold the edge of the cover glass between the thumb and index finger of your right hand and scrape a very small amount of Vaseline from your palm. Rotate the cover glass and scape again until all four sides of the cover glass receive the Vaseline. Place the cover glass (Vaseline side facing up) on your desk.

4. Using a transfer loop, place a very small drop of the hay infusion on the center of the cover glass (Fig. 6-1b).

5. Invert the hanging drop slide (depression side toward bench), carefully lowering it onto the cover glass until contact with the Vaseline is made. The

PREPARATION OF THE HANGING DROP

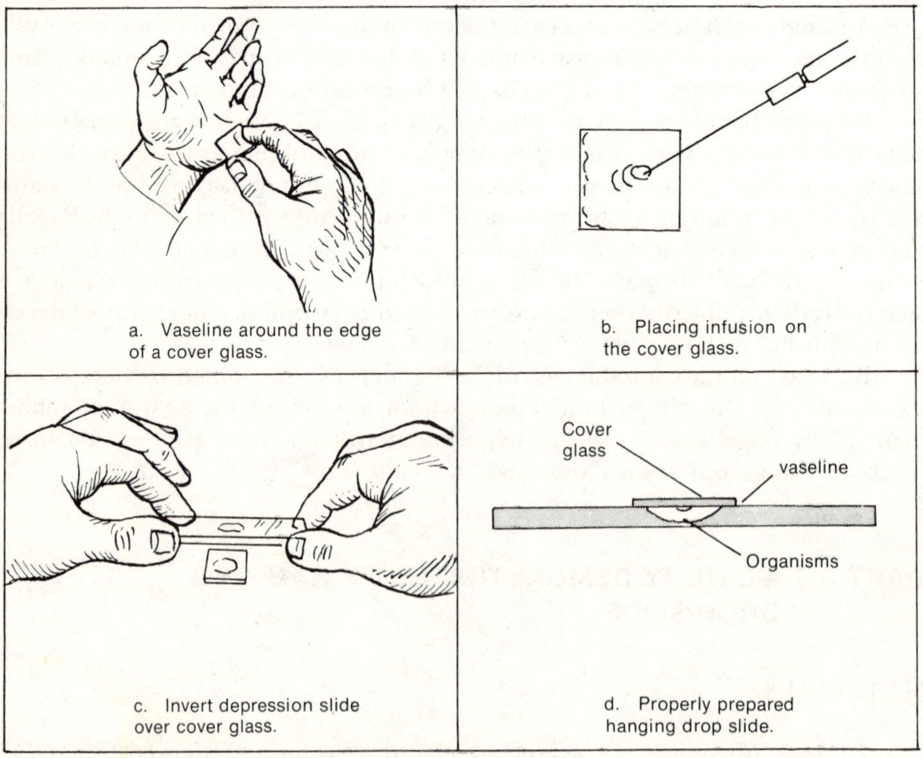

a. Vaseline around the edge of a cover glass.

b. Placing infusion on the cover glass.

c. Invert depression slide over cover glass.

d. Properly prepared hanging drop slide.

FIGURE 6–1

Vaseline around the edge of the cover glass will spread out and act as a seal. Pick up the slide and turn it right side up. In properly prepared slides, the drop of hay infusion remains suspended from the cover glass.

6. Examine the hanging drop with the low power objective. Close the iris diaphragm down to a small opening to improve contrast. Focus on the very edge of the drop. Can you see large protozoans rapidly moving about? If so, gently rotate the high dry objective into place, focus and admit slightly more light.

7. While focused on the very edge of the drop, you should be able to see some of the bacteria present, but they will appear extremely small. Rotate the high dry objective aside, place a drop of oil on the cover glass, and then examine with the oil immersion objective. Sharpen your focus and admit more light (don't increase intensity enough to destroy contrast).

8. Note that motile bacteria seem to have a directed motion while nonmotile cells stay in the same general area and show only Brownian movement.

9. Using aseptic techniques, make hanging drop preparations for each of the three pure cultures and examine under the oil immersion objective. Always start with the low power objective and find the edge of the drop. *Proteus vulgaris* is motile. You will determine whether the other two cultures are motile. Record your results in the space provided.

PART B. MOTILITY DEMONSTRATED BY WET MOUNTS

A wet mount is prepared by placing a drop of culture on a slide and covering it with a cover glass. Since the cover glass is not sealed, evaporation takes place along the edges and causes the suspended microorganisms to flow toward the edges of the wet mount. Thus, one must work quickly before the specimens dry out and examination should be limited to the center of the wet mount.

MATERIALS

1. Slides and cover glasses.
2. Pure cultures used in Part A.

PROCEDURE

1. Using aseptic techniques, transfer a loopful of *Proteus vulgaris* to the center of a clean slide. Cover with a clean cover glass and examine the center of the wet mount with the oil immersion objective.

2. Be careful not to crush the cover glass with the objective and keep light intensity low.

3. Observe culture for motility, working quickly. Repeat the procedure for the other two cultures. Record motility in the space provided.

COMMON AQUATIC ORGANISMS: PIGMENTED AND NONPIGMENTED, FLAGELLATE AND CILIATE PROTOZOANS

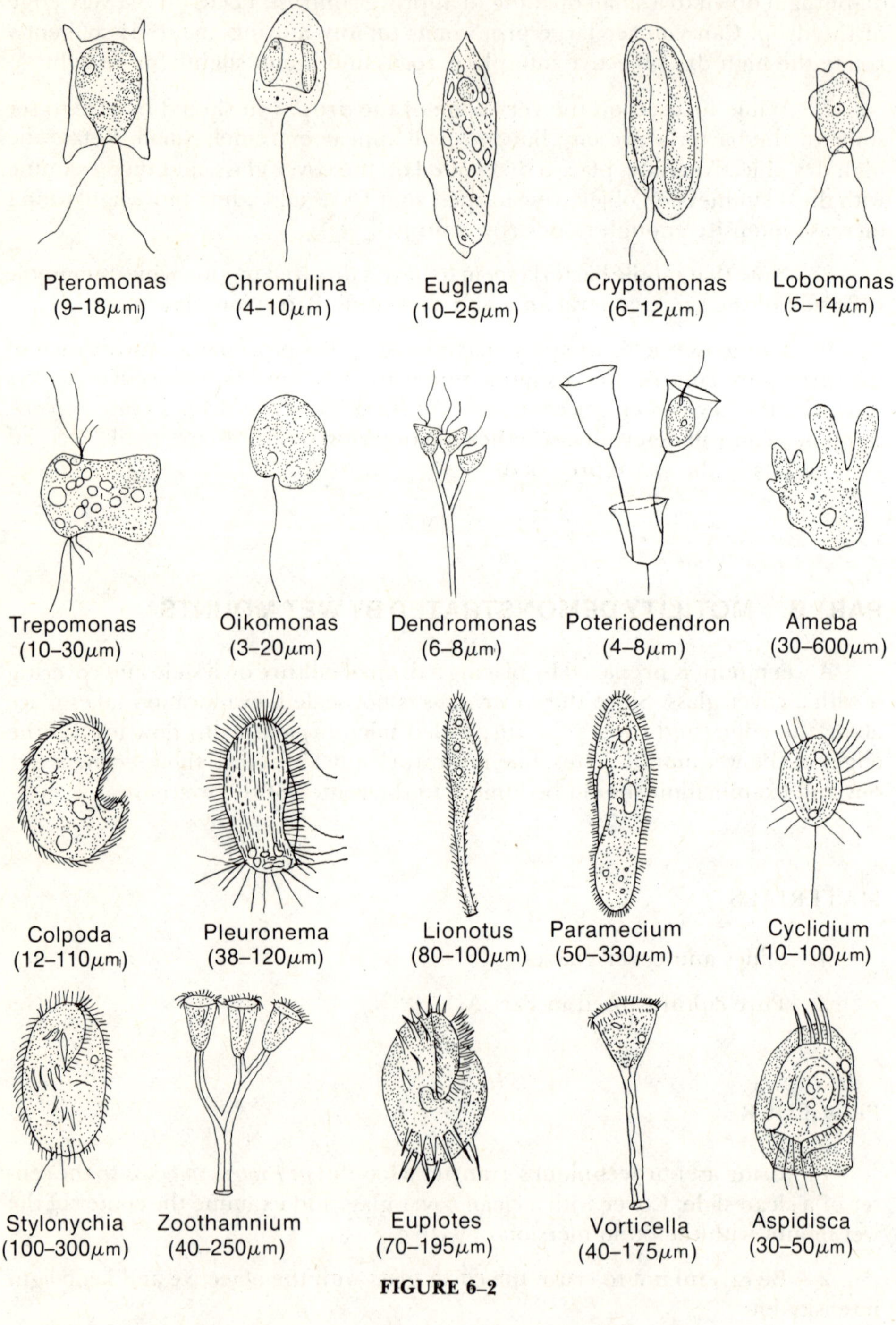

FIGURE 6–2

(Modified from *Standard Methods for the Examination of Water and Wastewater,* 13th ed. New York: American Public Health Association, 1971.)

AQUATIC ORGANISMS: COMMON BLUE-GREEN AND GREEN ALGAE

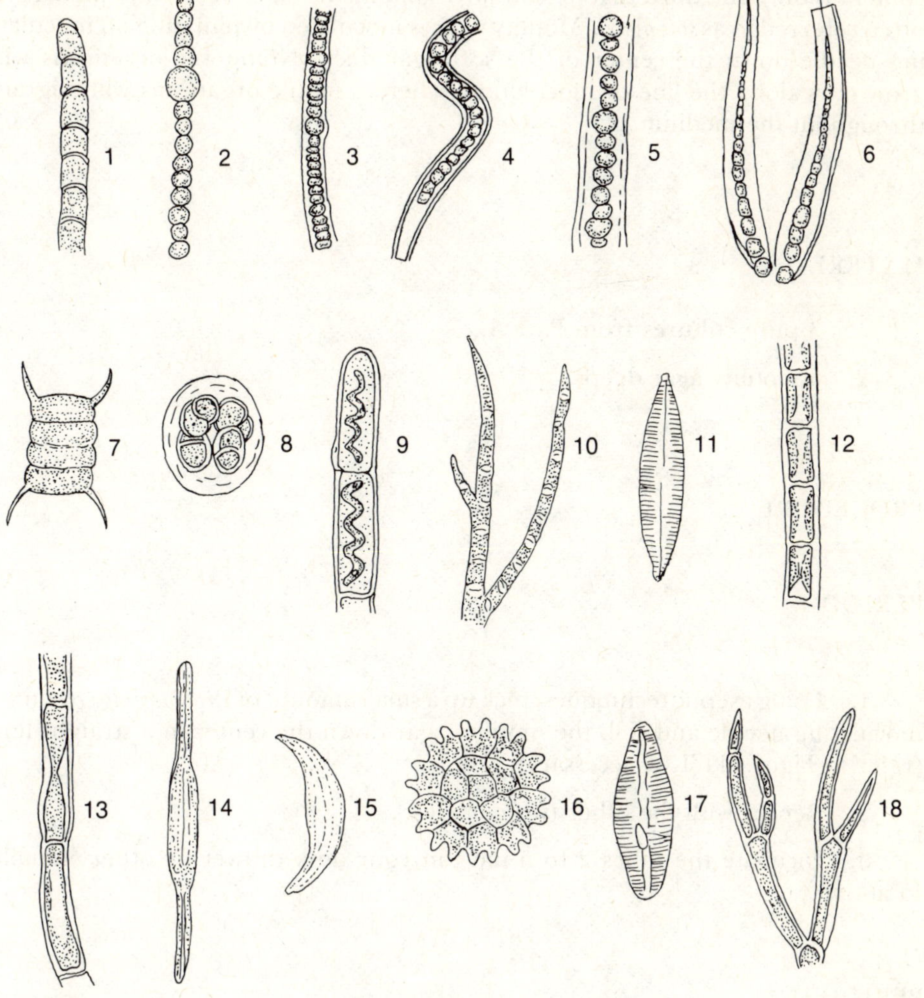

FIGURE 6–3.

1. Oscillatoria (4–20)
2. Anabaena (5–12)
3. Nodularia (5–10)
4. Lyngbya (4–20)
5. Cylindrospermum (5–10)
6. Rivularia (4–10)
7. Scenedesmus (4–6)**
8. Oocystis (2–6)
9. Spirogyra (15–100)
10. Stigeoclonium (2–10)
11. Navicula (20–30)*
12. Ulothrix (10–50)
13. Mougeotia (10–50)
14. Fragilaria (70–100)*
15. Closterium (10–15)
16. Pediastrum (10–20)**
17. Diatoma (15–20)*
18. Cladophora (4–20)

*Dimensions refer to length of cells in μm.
**Dimensions refer to individual cells.

All other dimensions refer to diameter of individual filaments or to mass.

(Modified from *Standard Methods for the Examination of Water and Wastewater,* 13th ed. New York: American Public Health Association, 1971.)

PART C. MOTILITY DEMONSTRATED BY MOTILITY AGAR

Motility agar is a special medium containing 0.2 per cent glucose, 0.3 per cent peptone, 0.2 per cent yeast extract, and 0.5 per cent agar. Note that the agar content is only one third that of standard solid media; such semi-solid media are often referred to as *soft agars.* Motility agar is inoculated by plunging an inoculating needle down the center of the soft agar deep. Nonmotile organisms will grow only along the line of inoculation, whereas motile organisms will migrate throughout the medium.

MATERIALS

1. 3 pure cultures from Part A.
2. 3 motility agar deeps.

PROCEDURE

PERIOD 1

1. Using aseptic techniques, pick up a small amount of *Proteus vulgaris* on an inoculating needle and stab the motility agar down the center in a straight line (refer to Figure 11-1, if necessary).
2. Repeat with the other two cultures.
3. Incubate the tubes 2 to 5 days in your desk drawer or other suitable location.

PERIOD II

1. Examine all tubes to see whether the culture has grown only along the line of inoculation or has spread throughout the soft agar.
2. Record your results in the space provided.

PART D. DEMONSTRATION OF FLAGELLA STAINS

Flagella staining is a difficult technique and beginning students generally lack the experience and time necessary to make good preparations. Therefore your instructor has prepared (or purchased) demonstration slides. Observe these stains and compare results obtained with the photomicrographs of flagellated bacteria also on display.

Students wishing to make flagella stains as a special project are encouraged to do so. See H. J. Conn and M. W. Jennison, eds., *Manual of Microbiological Methods* (McGraw-Hill Book Co., Inc.: New York), 1957, pp. 24 to 28, for detailed instructions.

S non motile

EXPERIMENT 7

Survey of Higher Protists: Actinomycetes, Yeasts, and Molds

The organisms in the Order Actinomycetales are procaryotic and are considered "higher" bacteria. However, for a long time these organisms were regarded as "ray fungi" because of their morphological similarity to many of the molds.

The majority of the organisms in the Order are saprophytic and aerobic, but some are parasitic and strictly anaerobic. A few produce human and animal diseases: tuberculosis *(Mycobacterium tuberculosis),* leprosy *(M. leprae),* actinomycosis *(Actinomyces israelii),* nocardiosis *(Nocardia asteroides),* and dermatitis *(Dermatophilus congolensis).* However, many species in the Families Streptomycetaceae and Micromonosporaceae are sources of such important antibiotics as streptomycin and tetracycline.

The organisms belonging to the Family Mycobacteriaceae do not form a *mycelium* (a mass of thread-like filaments or *hyphae*) or spores and often display false branching of the filaments, which are composed of long chains of individual rod-shaped cells. Those organisms producing various kinds of spores as well as a mycelium (that subsequently disintegrates into rods or cocci) and displaying true branching belong to the Family Actinomycetaceae. The Family Streptomycetaceae consists principally of soil bacteria. The mycelium does not break up into individual rods or coccoid forms. The Streptomycetaceae form reproductive bodies known as *conidia* (spores in chains) on special spore-bearing hyphae (Fig. 7-1).

The fungi (yeasts and molds) are generally classified on the basis of sexual reproduction and morphology. A simplified and highly abbreviated scheme appears below. Further details will be found in your text.

1. Phycomycetes	Essentially nonseptate hyphae, asexual spores in sporangium, sexual reproduction by conjugation of isogametes or by production of sexual organs with dissimilar gametes. Examples: *Mucor* sp. or *Rhizopus* sp.
2. Ascomycetes	Septate hyphae, sexual spores in ascus or sac, asexual spores exogenous or free. Examples: *Saccharomyces cerevisiae, Penicillium* sp., *Aspergillus* sp.
3. Basidiomycetes	Septate hyphae, sexual spores borne exogenously on basidium. Examples: *Agaricus* sp., mushrooms, puffballs.
4. Deuteromycetes (Fungi imperfecti)	Septate hyphae, no sexual stage observed, asexual spores exogenous. Examples: *Histoplasma capsulatum* and *Candida albicans.*

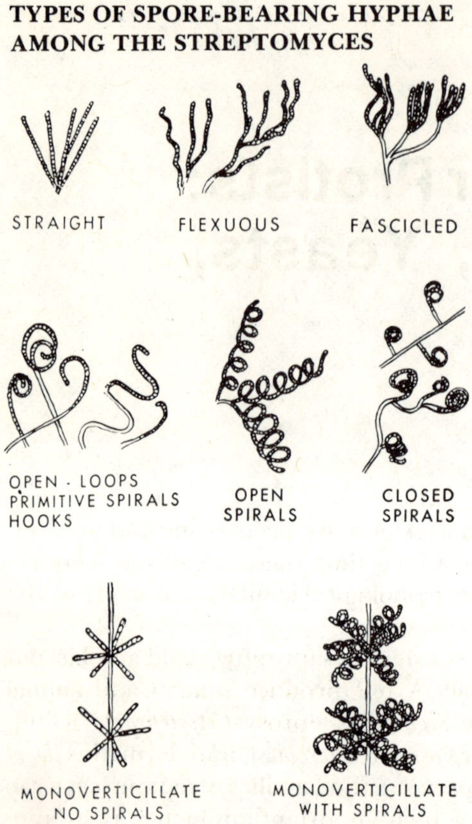

FIGURE 7—1. (From Frobisher, M., et al. *Fundamentals of Microbiology,* 9th ed. Philadelphia: W. B. Saunders Co., 1974.)

Yeasts, eucaryotic organisms, comprise two of the above classes: *Ascomycetes* and Deuteromycetes (Fungi imperfecti). Yeasts belonging to the Class Ascomycetes have a perfect (sexually reproductive) stage, forming *ascospores* (spores derived from fusion of + and − strains of yeast cells). These yeasts are also known as "true" yeasts; however, they usually multiply by asexual means, that is, by budding. Yeasts classified as Deuteromycetes do not possess a perfect or sexual stage and are often called "false" yeasts because they do not form ascospores. Their reproduction is limited to the asexual processes of budding and fission (Fig. 7-2a). While some yeasts *(Candida albicans)* are pathogenic to man and other animals, many *(Saccharomyces* sp.) are important in such varied industrial processes as alcoholic fermentation, fat production, and baking.

Molds are eucaryotic and are usually considered to be multicellular in their development, forming hyphae and mycelia. The structures bearing the asexual spores and the presence or absence of a *septum* (cross wall) in vegetative hyphae are important morphological characteristics used for identification and classification. For example, vegetative hyphae of bread mold *(Rhizopus nigricans)* lack septa and the asexual spores are borne on a structure known as the *columella.* Both columella and spores are enclosed in a spore-bearing structure termed the *sporangium.* Because of this, and its mode of sexual reproduction, the bread mold belongs to Class Phycomycetes (algal or water molds, Fig. 7-2c). In contrast, vegetative hyphae of the common black mold (*Aspergillus niger,* Fig. 7-2e) and the green mold (*Penicillium chrysogenum,* Fig. 7-2d) are septate. The spores of *Aspergillus* are borne on flask-shaped structures called *sterigmata* (singular,

VEGETATIVE STRUCTURES OF SOME TYPICAL FUNGI: YEAST AND MOLDS

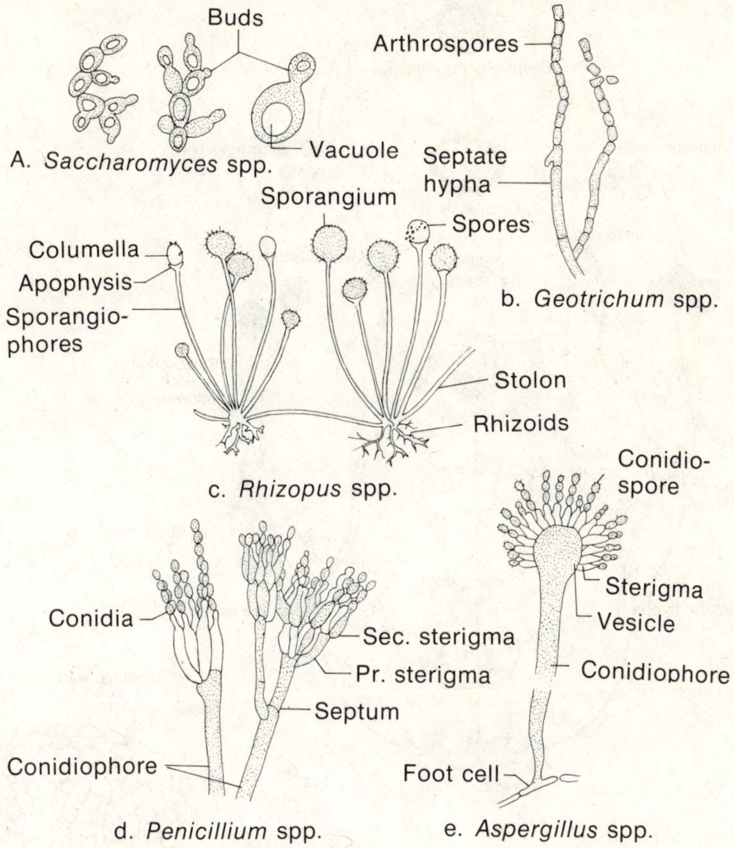

FIGURE 7–2

sterigma) which are attached to the *vesicle,* the swollen end of the *conidiophore* (conidia-bearing structure). The conidiophore arises from the vegetative hyphae.

Both Phycomycetes (nonseptate, filamentous molds) and Ascomycetes (e.g., septate molds such as *Aspergillus* and *Penicillium*) have a perfect stage. Fusion of + and − hyphae forms a diploid structure known as a zygospore *(Rhizopus,* Fig. 7-3a) or ascospore *(Aspergillus* and *Penicillium).* Zygospores will eventually germinate to produce sporangiospores of either + or − strain. Ascospores (Fig. 7-3b) are borne in a structure known as an *ascus* (sac), which is frequently surrounded by a mass of mycelia known as the *ascocarp.* For this reason the latter two organisms are termed Ascomycetes, or sac fungi. Students are encouraged to refer to the text for detailed reproductive cycles.

Only a few molds are pathogenic to man and other animals, the great majority of them being used in various industrial processes, such as production of organic acids, alcohols, solvents, and antibiotics (e.g., penicillin).

MATERIALS

1. Pure cultures of *Streptomyces griseus* (common soil actinomycetes) on glucose yeast extract agar plates; pure cultures of *Saccharomyces cerevisiae* (budding

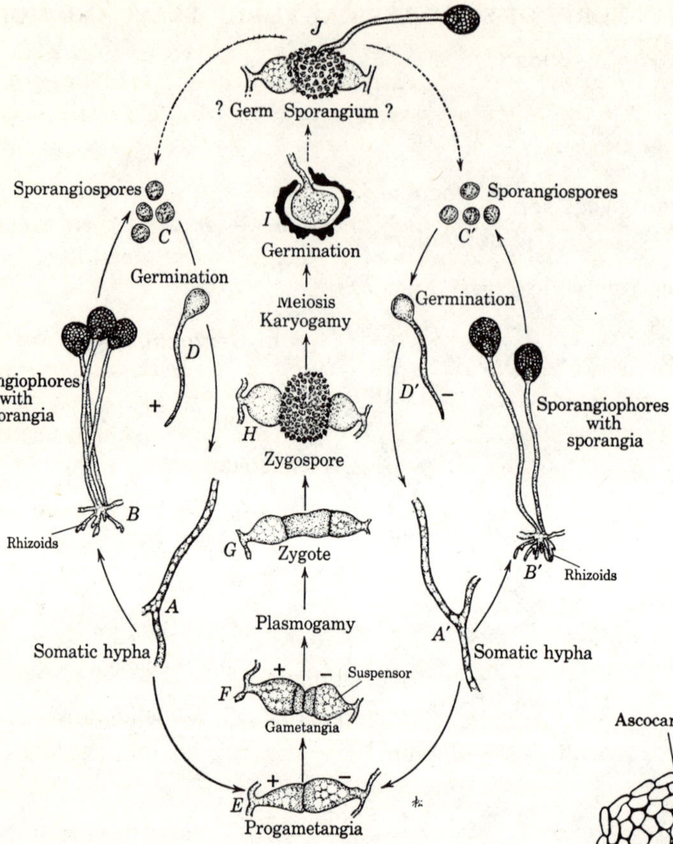

A

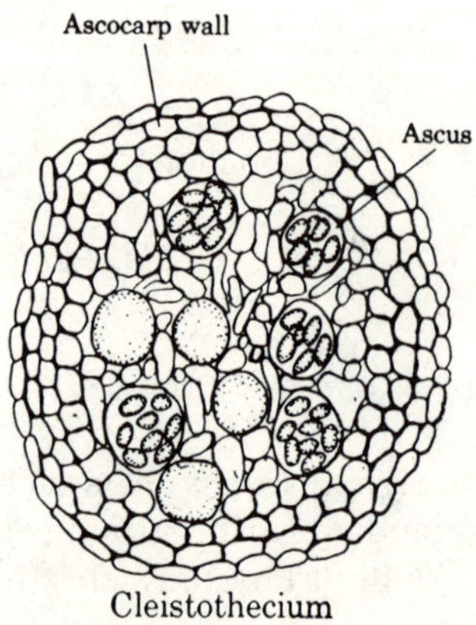

B

FIGURE 7–3a. Life cycle of Phycomycetes *(Rhizopus nigricans).*

FIGURE 7–3b. Sexual reproductive body *(fruiting body)* of Ascomycetes.

(From Alexopoulos, C. J. *Introductory Mycology,* 2nd ed. New York: John Wiley and Sons, Inc., 1962. 7–3a redrawn from Cutter, *Bull. Torrey Bot. Club, 69*:592–616, 1942.)

yeast), *Geotrichum candidum* (yeast-like mold or "dairy mold"). *Rhizopus nigricans* (common bread mold), *Aspergillus niger* (soil black mold), *Penicillium chrysogenum* (green mold) on Sabouraud dextrose agar plates. (See Appendix for formulas.)

PROCEDURE

1. Observe colonies on the agar plates (use low power stereoscope, if available) and compare color and types of growth for all the organisms. Record your observations in the spaces provided.

2. *Streptomyces griseus:* Using a sterile culture loop, scrape off a small amount of the growth and make a wet mount of the specimen. Place a cover glass over the preparation and observe with oil immersion objective. Draw the structure of the hyphae. Are the hyphae nonseptate? Compare the size of the streptomyces spore with an endospore.

3. Examine the plate culture of *Saccharomyces cerevisiae* and record the gross appearance of the colony. Remove a small amount of the cells from the edge of the colony and make a wet mount. Observe with the oil immersion objective. Record your observations in the space provided.

4. *Geotrichum candidum (Oöspora lactis):* Prepare a wet mount as above. Observe the dichotomous branching and septate hyphae. The asexual spores are arthrospores *(oidia)* which may appear rectangular if the growth is taken from submerged hyphae (embedded into the agar) and oval if taken from aerial hyphae (white powdery area usually in the center of the colony).

5. *Rhizopus nigricans:* Using a sterile wire loop, obtain a fragment of the growth including fruiting body and prepare a wet mount. Observe and note virtual absence of cross wall in vegetative hyphae. Brownish root-like structures on vegetative hyphae are called *rhizoids* and are usually anchored into the agar medium. The growing tip of vegetative hyphae is called the *stolon.* Locate and draw the *sporangiophore, sporangium,* and *columella* of a ruptured sporangium.

6. *Aspergillus niger:* Repeat Step 5 for this microorganism. Observe and draw vegetative hyphae, foot cell, conidiophore (note conidiophore in this organism is nonseptate), vesicle, sterigmata, and conidia.

7. *Penicillium chrysogenum:* Do the same as above. Note septate conidiophore and absence of vesicles. Draw and record all of your observations in the space provided.

EXPERIMENT 8

Culture Media Preparation and Sterilization

The nutritional requirements of bacteria are extremely variable, although bacteria, as living things, have certain basic needs in common. In addition to water (H_2O), four essential elements, carbon (C), nitrogen (N), phosphorus (P), and sulfur (S) must be present in the media. These four elements may be provided by inorganic or organic compounds, depending upon the kind of organism one wishes to cultivate. For example, *photoautotrophs* and *chemoautotrophs* require the necessary elements in inorganic form as they are capable of synthesizing all of their cellular constituents by using either sunlight or the oxidation of inorganic compounds as a source of energy. All organisms also require vitamins and trace mineral elements such as magnesium (Mg), zinc (Zn), copper (Cu), and molybdenum (Mo). Autotrophs can synthesize their own vitamins, while many heterotrophs require preformed vitamins (or precursors of vitamins) as they are unable to synthesize their own.

A culture medium is the material in which or on which bacteria are grown. The media used for the cultivation of microorganisms may be liquid, semisolid, or solid. The ingredients included in any one medium may be limited to well-defined inorganic or organic compounds. A medium such as this, whose chemical constituents are well known, is said to be a *chemically defined medium.* On the other hand, media (especially those designed for cultivation of nutritionally fastidious organisms, such as certain human pathogens) may contain exceedingly complex materials (digest of plants, animal tissue extract, or blood serum) whose composition is not clearly defined.

Nutrient broth, or nutrient agar, is the medium most commonly used in bacteriology. The medium supports the growth of a large number of organisms, hence it is called *all purpose* or *general purpose medium,* in contrast to *selective* or *differential* media. Selective media are media containing chemical substances which will prevent growth of one or more groups of bacteria without inhibiting the growth of the desired one. Differential media contain certain chemicals which permit the observer to differentiate between types of bacteria. Representatives of these media types will be used in later experiments.

Nutrient broth consists of beef extract (3 grams per liter) and peptone, partially digested beef protein (5 g/l), dissolved in distilled water. (Your instructor will explain how these two ingredients are prepared from raw materials.) If a solid medium is desired, 15 g/l of agar may be added to the broth medium. Although the primary constituent of agar is a complex carbohydrate, galactan, extracted from the marine alga of the genus *Gelidium,* it is not used as food by the vast majority of microorganisms but acts only as a solidifying agent. Agar goes into solution or melts when heated to nearly 100C and remains liquid until cooled to about 43C. Unlike gelatin, once solidified it must be reheated to approximately 100C to cause liquefaction.

In bacteriology, the terms "sterile" and "sterilized" mean free of all life, including viruses. Culture media, with the exception of a few, must be sterilized in proper containers immediately after preparation. The sterilized medium must be kept in sterile condition if it is to be of any service in bacteriological studies. Remember, bacteria and molds are always present in the water and on glassware and other utensils. Storing unsterile media in a refrigerator merely delays the growth of the organisms.

The principal lethal agent in steam sterilization is heat. Microorganisms are killed when their enzymes and cellular proteins are irreversibly destroyed. The higher the temperature, the shorter the time required for sterilization of the media.

The *autoclave* is an instrument used to sterilize bacteriological media and surgical equipment (see Fig. 8-1). It is based on the same principle as the pressure cooker: the ability to increase temperature by building up steam pressure, thus preventing boiling. In normal steam sterilization, the temperature is kept at 121C (at this temperature the pressure gauge should read between 15 and 16 pounds per square inch) for 15 to 20 minutes. Most media and equipment used in bacteriology are sterilized by this process. Although a higher temperature can be obtained by increasing the steam pressure, further increase in temperature and time may drastically alter the ingredients in the media, thereby making them unsuitable for the particular use.

Use of the autoclave will be demonstrated by your instructor. Keep notes as to loading, steam generation, pressure gauge, timer, and other pertinent information, particularly with regard to the *safe* operation of autoclaves.

MATERIALS

1. Bacto-peptone; beef extract; agar.

2. Trip balance for each group.

3. 1-6 oz prescription bottle with cap per student; 1-1000 ml Erlenmeyer flask per group.

4. 3 nonsterile test tubes per student; wads of plugging cotton; applicator sticks. (Your instructor will show you how to make cotton plugs and cotton swabs.)

5. 1-50 ml, 1-100 ml, and 1-1000 ml graduated cylinder per group.

PROCEDURE

PERIOD I

1. A group of eight students will work together to prepare the medium.
2. Add 1000 ml of distilled water into a liter Erlenmeyer flask.

PRESSURE STEAM STERILIZER (AUTOCLAVE) AND HOT AIR STERILIZER (OVEN)

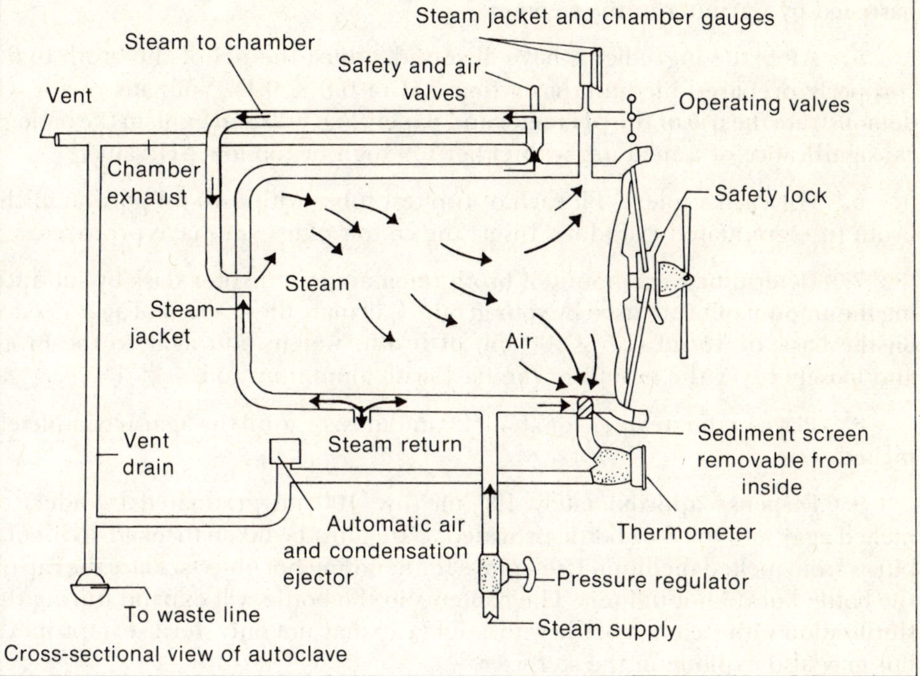

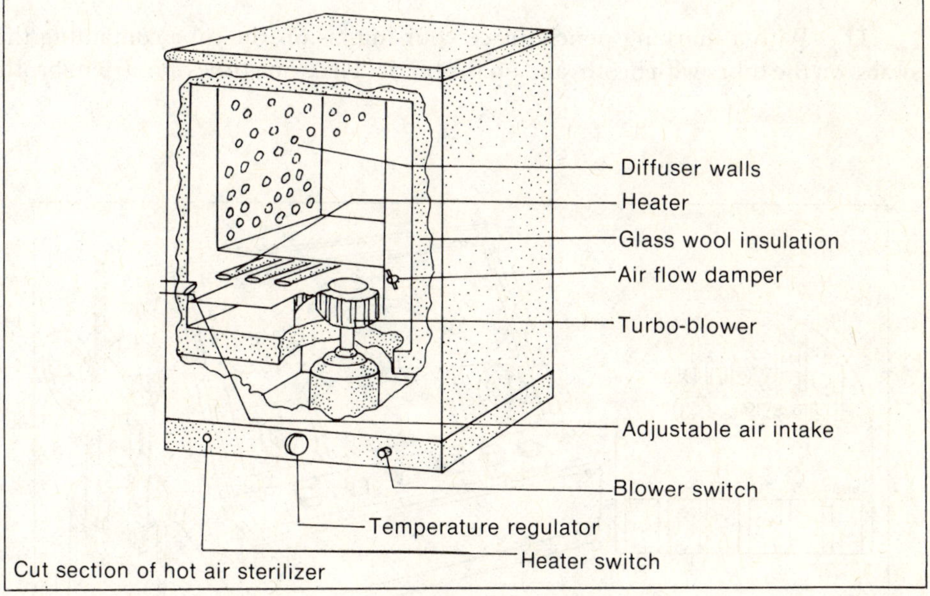

FIGURE 8–1

3. Weigh 5 g of peptone and put it into the flask.

4. Weigh 3 g of beef extract and put it into the flask. Since beef extract is rather sticky, it should be weighed on bibulous paper. Both the beef extract and paper can be placed in the water until the beef extract slips off, and the paper can be removed with a wire loop. Solubilization of ingredients may be hastened by warming in the steamer.

5. After the ingredients have dissolved, adjust the pH of the broth to 6.8 (properly prepared medium has a final pH of 6.8 ± 0.1). Your instructor will demonstrate the use of the pH meter and paper (Fig. 8-2) and explain the biological significance of a medium with either too high or too low pH value.

6. For each student, fill each of two test tubes with a 10 ml portion of the broth (use graduated cylinder). Insert the cotton plugs you have prepared.

7. Determine the amount of broth remaining in the liter flask by subtracting the amount of broth used by your group. Calculate the amount of agar needed on the basis of 15 g/l (1.5 g/100 ml) of broth; weigh; add agar to the broth and loosely cover the mouth of the flask with aluminum foil.

8. Place in the steamer for about 20 minutes, or until the agar is completely melted.

9. Dispense approximately 100 ml (use 100 ml graduated cylinder) of melted agar into the 6 oz bottle provided. Care must be taken to avoid accidental burns from melted medium. Use a glove for handling hot objects. Place the cap on the bottle but do not tighten. The contents in the bottle will expand during the sterilization with steam; a bottle with a tight cap may not only sterilize improperly but may also explode in the sterilizer.

10. Place the cotton swabs you have prepared in a test tube. Insert a cotton plug.

11. With a marking pencil, place your initials on the tube containing the swabs, on the tubes with broth, and on the bottle. Place one tube of nutrient broth,

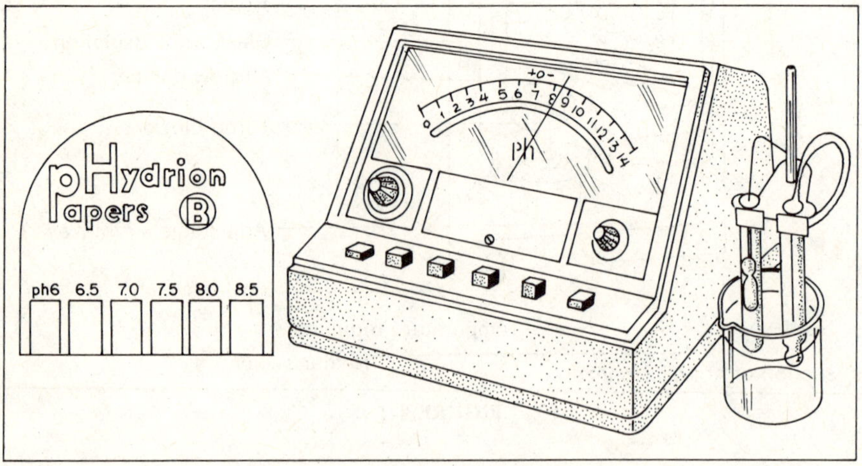

FIGURE 8–2. pH paper and meter.

the tube with swabs, and the bottle of agar into the receptacle provided for sterilization. The sterilized media will be returned to you at the next laboratory period.

12. Keep the second tube of broth in your desk cabinet and incubate it until the next laboratory period. Remember the medium is unsterile.

PERIOD II

1. Compare sterilized nutrient broth with the unsterile broth you incubated in your desk cabinet. Record your observations for turbidity (cloudiness) in the space provided.

2. Mix the contaminated broth. Make a smear; air dry, heat fix, and stain with aqueous methylene blue for 2 to 5 minutes. Observe the specimen under oil immersion and sketch your observations in the space provided.

RESULTS

Smear from nonsterile medium

Description of nonsterile medium

DISCUSSION

1. What is the reason for raising the internal pressure of the autoclave to 15 to 16 pounds per square inch (psi)?

2. How and from where do you suppose the bacteria which grew in the unsterile medium were introduced?

EXPERIMENT 9

Environmental Distribution and Enumeration of Microorganisms

While requirements for survival of organisms are not as rigid as those for growth, certain conditions must be met if an environment is to serve as a habitat for microorganisms. In natural environments one generally finds adequate supplies of water and food materials, and suitable pH, oxygen tension, and temperature for microbial growth. Thus it is known that microorganisms are present in or on almost everything: soil, water, air, plants, animals, in the intestinal tract. Some organisms are indigenous to a certain environment; others may be only transient. For example, a group of organisms known as coliforms are always harbored in the intestinal tract of man and other warm-blooded animals. In contrast, the microorganisms found in air differ markedly from region to region because air alone does not support the growth of microorganisms. The "microflora" of air are usually soil- and water-borne organisms, in a state of suspended animation until they are washed down by rainwater. Because the numbers and kinds of microorganisms depend upon so many factors, the populations in any given environment are usually expressed in ranges (i.e., 1×10^5 to 1×10^9 cells per gram of soil) rather than in absolute numbers.

The presence of microorganisms in a fertile soil may be demonstrated by direct microscopic examination or by plate count. It is quite impossible to demonstrate the presence of microorganisms in air by the same method because of the small numbers present. Nevertheless, their presence can be readily ascertained by exposing sterile media to air. Likewise, the presence of microorganisms in or on "things" can be demonstrated by either inoculating the "thing" into sterile medium or by removing the organisms with sterile swabs and subsequently culturing the organisms present on the swabs in suitable medium.

Unlike direct microscopic observation, the plating method requires only a small number of organisms for examination. If the number of organisms inoculated per plate is too great, the entire plate is covered by the growth (confluent growth) and individual colonies cannot be observed. In the plating method, the number of organisms in the sample is reduced by a method known as serial dilution. The series of dilutions is usually made in multiples of 10 or 100, that is, 1 g or 1 ml of sample is added to 9 ml of a sterile water blank (1:10), or to 99 ml of the blank (1:100). The dilution required depends upon the number of

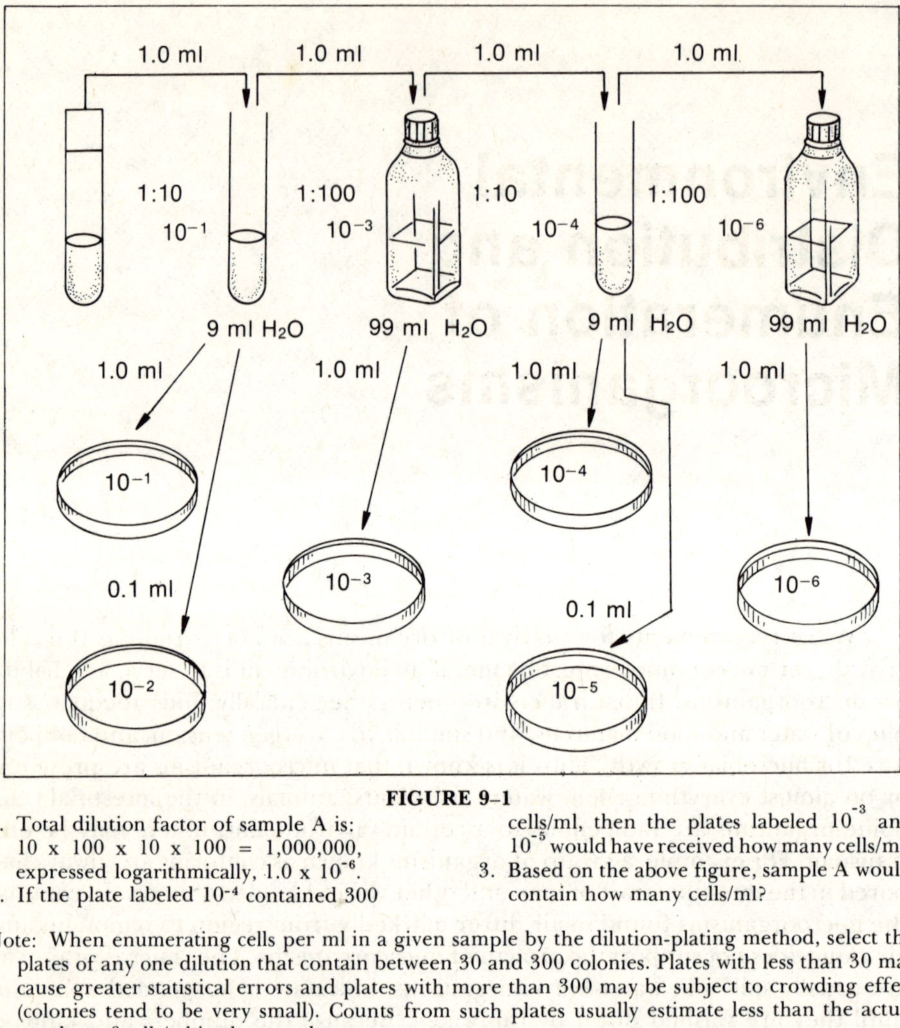

FIGURE 9–1

1. Total dilution factor of sample A is: 10 x 100 x 10 x 100 = 1,000,000, expressed logarithmically, 1.0×10^{-6}.
2. If the plate labeled 10^{-4} contained 300 cells/ml, then the plates labeled 10^{-3} and 10^{-5} would have received how many cells/ml?
3. Based on the above figure, sample A would contain how many cells/ml?

Note: When enumerating cells per ml in a given sample by the dilution-plating method, select the plates of any one dilution that contain between 30 and 300 colonies. Plates with less than 30 may cause greater statistical errors and plates with more than 300 may be subject to crowding effect (colonies tend to be very small). Counts from such plates usually estimate less than the actual number of cells/ml in the sample.

A standard practice is to prepare duplicate (ideally triplicate) plates for each dilution plated. The colony count should be the average of the two (or three) plates for that dilution. If the counts on duplicate plates vary from each other by much more than 10 to 15 per cent, the credibility of the count is lowered and your techniques are suspect.

microorganisms present in the initial sample (see Fig. 9-1). An *aliquot,* usually 1 ml of the diluted sample, is plated on suitable medium.

To demonstrate the presence of microorganisms in any environment, the media and equipment to be used must be sterile and the technique must be aseptic to prevent contamination from sources other than the one being investigated.

MATERIALS

1. 1 bottle of nutrient agar (from Experiment 8).
2. 1 tube of nutrient broth (from Experiment 8).
3. 5 sterile Petri dishes.

4. Samples of serially diluted soil (1:100,000).
5. Samples of serially diluted polluted water (1:1000).
6. Canister of pipettes.

PROCEDURE

PERIOD I

1. Label sterile Petri dishes as follows:

   a. On the lids of all Petri dishes (near periphery), clearly indicate date, Experiment number, Section number, and your initials. *Never* open the sterile dishes while labeling and handle carefully to minimize introduction of organisms other than those being investigated.

   b. Invert the dishes and number the bottoms as follows: (1) Soil, 1:100,000; (2) P. W. (polluted water), 1:1000; (3) Air, 10 min; (4) "things" (sector dish into 6 equal parts to accommodate inocula such as a bit of hair, a dime, fingertip print, etc.); (5) control. The control dish is to verify that the dishes and medium are sterile and that your technique is aseptic.

2. Obtain a sterile 1.0 ml pipette from a canister. Transfer 1.0 ml of the diluted soil sample into the dish so labeled. Discard the used pipette into the receptor provided. (Do not blow out the liquid from the pipette tip into the sink; the liquid may contain pathogens.)

NOTE: Expose sterile pipettes for a minimum period of time to prevent contamination from the air. Never put a sterile pipette on the bench or allow the tip of a pipette to come in contact with other objects before intended use. Never touch the tip of the pipette with your fingers. Your instructor will demonstrate the proper handling of pipettes and the aseptic method of inoculating a liquid sample into Petri dishes. See Figure 9-2.

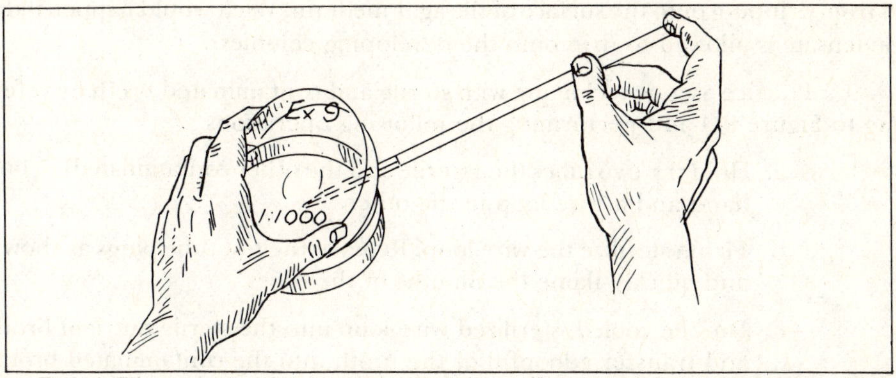

FIGURE 9–2. Proper holding of pipette and Petri dish for aseptic technique and control of sample volume.

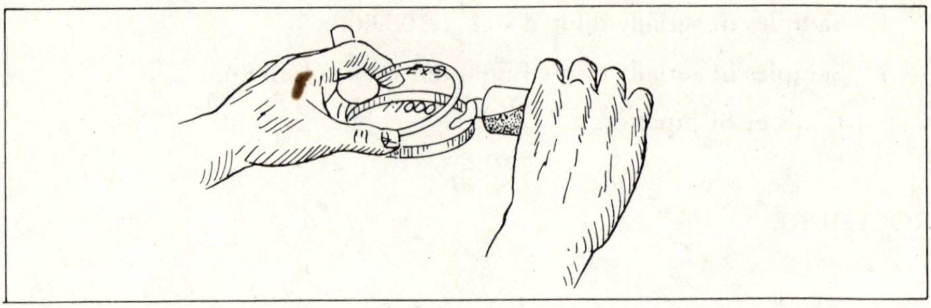

FIGURE 9–3. Aseptic technique for pouring plates.

3. Using a new pipette, transfer 1.0 ml of the diluted water into dish 2.

4. Obtain 1 bottle of melted, cooled (45C) agar from the water bath. Wipe off all the water on the exterior of the bottle with a paper towel. Why?

5. Arrange dishes 1, 2, 3, 4, and 5 in a row in front of you. After flaming the neck of the bottle, pour into each dish approximately 15 to 20 ml of agar, or enough to cover the bottom of the dish to the thickness of 2 to 3 mm (see Fig. 9-3). For dishes 1 and 2, the medium must be mixed thoroughly with the sample by tipping the dishes back and forth gently—this is to promote even distribution of the organisms. After mixing, allow the agar to solidify. Mixing is unnecessary for dishes 3, 4, and 5, as these dishes contain no inoculum.

6. After the agar has solidified (usually requires 5 to 10 minutes at room temperature; Petri dishes containing solidified agar are frequently referred to as Petri "plates"), remove the lid from plate 3 and expose to the air in the laboratory, or any place you desire, for 10 minutes. Replace the lid after exposure.

7. Inoculate plate 4 with various materials as described in Step 1b. Open the plate carefully. Raise one side just enough to allow insertion of the objects to be tested. The lid should be lowered immediately after each object has been placed in its proper sector. You may use a sterile swab to obtain mouth, nasal, or skin flora, if you wish, but it is strictly voluntary. Gently streak over the sector so labeled.

8. After the work has been completed, invert all the plates and incubate in your desk cabinet for 2 to 5 days. Inverting the plate prevents condensate on the lid from dripping onto the surface of the agar medium. What would happen if the condensate is allowed to drip onto the developing colonies?

9. Practice aseptic technique with sterile and contaminated broth by referring to Figure 9-4 and performing the following operations.

   a. Hold the two tubes (one sterile and the other contaminated) in one hand and a wire loop in the other.

   b. Flame sterilize the wire loop. Remove the test tube plugs as shown and quickly flame the mouths of the tubes.

   c. Dip the cooled, sterilized wire loop into the sterile nutrient broth and transfer a loopful of the broth into the contaminated broth. Flame the tubes, replace the plugs, and sterilize the loop again before you place it on the loop rest.

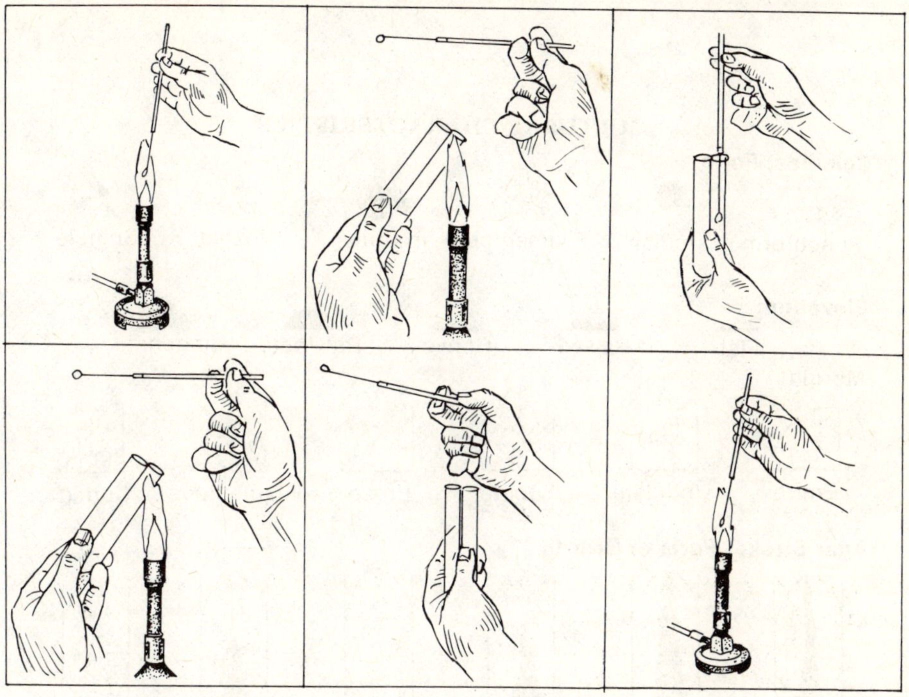

FIGURE 9–4. Aseptic handling of more than one tube at one time.

d. Repeat Steps a to c at least three times, using the same tubes, and incubate the broth tubes in your locker for 2 to 5 days.

Students who have practiced aseptic technique should anticipate no growth in the sterile nutrient broth as this tube has not come in contact with the contaminated material.

PERIOD II

1. Examine all of the Petri plates and record your observations as follows:
   a. Check your control plate and interpret the significance of the absence or presence of colonies on the agar. Also check your nutrient broth tube (from Step 9, Period I) for aseptic technique. If the tube labeled "sterile broth" is still completely clear, give yourself A for the aseptic handling of the media; if not, give yourself—.
   b. Examine the plates with soil and polluted water and estimate the approximate number of colonies in each plate (counts should include submerged colonies). The number of colonies in the plate multiplied by the dilution factor will give you the initial bacterial cell population in either a gram of the soil or a milliliter of polluted river water. *Do not* open the lid while counting. Observe diversity of colony color, shape, and pigments. Proper terms describing colony form, elevation, and margin are presented in Figure 9-5.

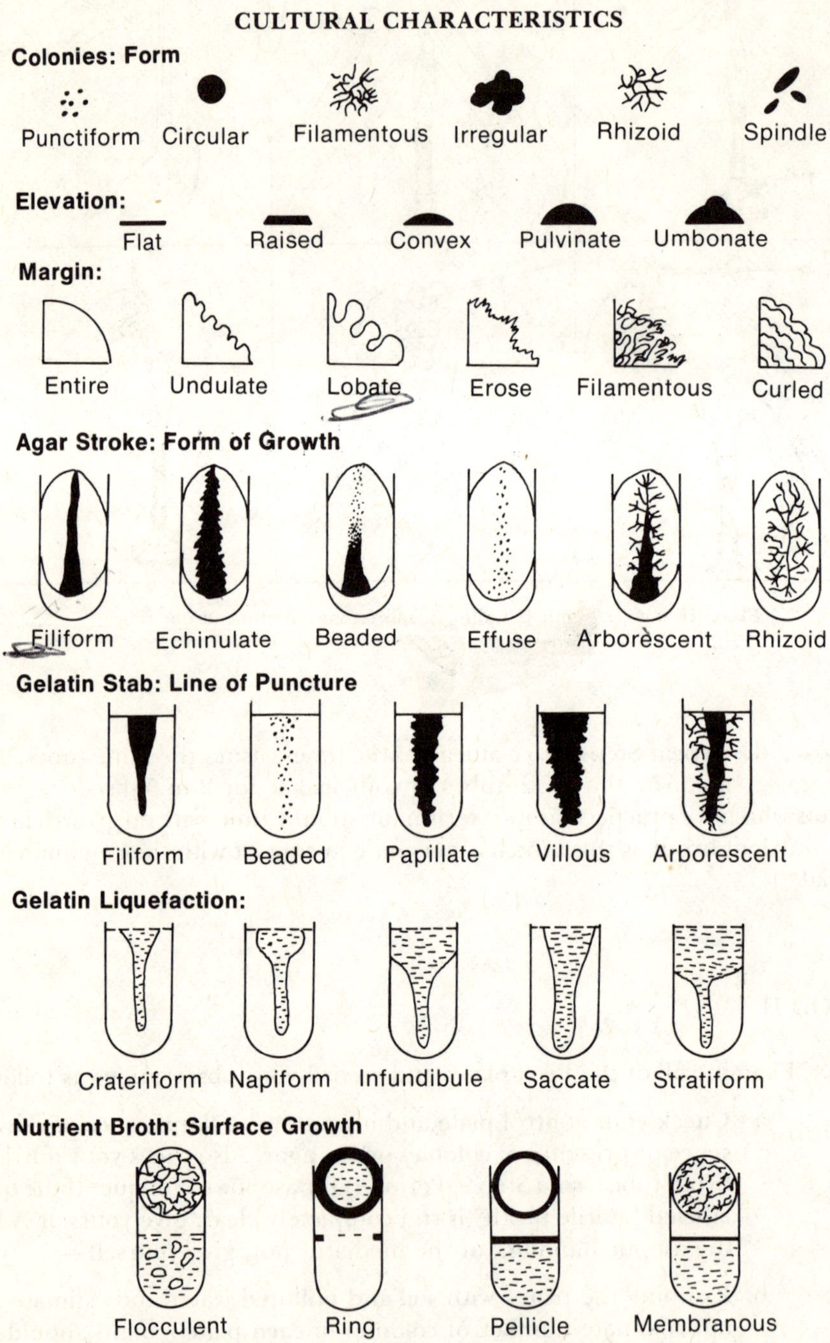

FIGURE 9–5

(Adapted from Conn, H. J., and Jennison, M. W., eds. *Manual of Microbiological Methods.* New York: McGraw-Hill, 1957.)

c. Examine the sectored plates. Were there any bacteria on your fingertip? On a dime? Record your observations in the space provided.

d. After you have completed your examination of the colonies on the plates, select any one colony well separated from all other colonies. With a sterile wire needle, transfer a small inoculum of cells from this colony into the tube of nutrient broth which you have proven sterile. (If contaminated, ask for replacement.) Mix the contents of the tube and incubate in your desk cabinet until the next period. Flame your wire needle before putting it away.

A bacterial colony, when it is well separated from others, with uniform shape, color, and texture, may be pure; that is, it may be a clone of organisms all the cells of which are the same kind and derived from a single parent cell placed on the agar surface. Such colonies will be used in the next experiment. Keep all the plates until next period.

RESULTS

1. Is there any microbial growth on the control plate or in the nutrient broth? If your answer is yes, explain briefly.

2. Record the initial microbial population in 1 g of soil and 1 ml of polluted river water. Use exponents to express the numbers.

____________________cells/g of soil

____________________cells/ml of polluted water

3. Write word descriptions of representative colonies which developed in plates inoculated with the soil and water.

4. If "things" (fomites, food, hair, etc.) inoculated on plate 4 were contaminated with microorganisms, describe the growth below:

Sector 1

Sector 2

Sector 3

Sector 4

Sector 5

Sector 6

DISCUSSION

1. The student is advised to reread the note which appears on page 74. Calculate the total number of cells/ml on the basis of the following information:

	Colonies	
Dilution	*Plate I*	*Plate II*
10^{-3}	TNC*	TNC*
10^{-4}	420	440
10^{-5}	43	49

*TNC = too numerous to count.

2. Describe the following:

   a. Colony elevation—"pulvinate":

   b. Colony margin—"entire":

   c. Submerged colony form—"spindle":

EXPERIMENT 10

Pure Culture Techniques

PART A. PURIFICATION OF A CULTURE BY DILUTION AND PLATING

A pure culture contains a single kind of microorganism. Only with a pure culture of a given species are the following studies possible: (1) *morphologic* (shape and size of the cells); (2) *cultural* (growth characteristics of the culture on various media); (3) *physiological* or *biochemical* (chemical changes in the medium caused by the culture); (4) *pathogenic* (ability of the organism to cause disease in plants or animals); (5) *serologic* (study of cellular antigens and serum antibodies); (6) *genetic* (variation, sexuality, DNA base ratio of the culture, DNA-DNA, DNA-RNA hybridization); and (7) *bacteriophage type* (the cells subject to lysis by specific bacteriovirus(es).

Although pure cultures are found in nature, the frequency with which they occur in any given environment is still considered low. A small crumb of fertile soil may contain a considerable array of viruses, heterotrophic and autotrophic bacteria, and many species of yeasts, molds, algae, and protozoans. These organisms may exist as saprophytes, commensals, symbionts, or antagonists.

To isolate a pure culture of a desired microorganism may require the use of special kinds of media. Three general types of media are frequently used: (1) *enrichment,* which contains substances favoring the growth of the particular organism being sought; (2) *selective,* which inhibits the growth of unwanted organisms but does not inhibit the growth of the type being sought; (3) *differential,* which contains ingredients that allow one to distinguish between two or more types of microorganisms (blood agar plates can be used to distinguish between bacteria producing different types of hemolysins; a pH indicator added to a medium will allow a microorganism that can ferment the sugar present to be distinguished from one that cannot). While there are media which serve each of the above purposes separately, students should also realize that certain types of media (e.g., eosin methylene blue agar—EMB) contain ingredients which act simultaneously to enrich, select, and differentiate.

The enrichment and selective agents may be chemical or physical. To isolate thermophilic bacteria, one incubates inoculated media at an elevated temperature (50C or higher) to discourage all other types. To isolate halophilic (salt-loving) organisms, one usually adds 10 per cent, or greater, sodium chloride (NaCl) to the medium; to isolate osmophilic bacteria, one frequently adds a high concentration of sugar to the medium. Many unwanted bacteria may undergo plasmolysis (shrinkage of the cytoplasm) in such media while desired ones are unaffected. Adjusting the pH of the medium high or low, adding specific antibiotics

LOOP DILUTION—POUR PLATE TECHNIQUE

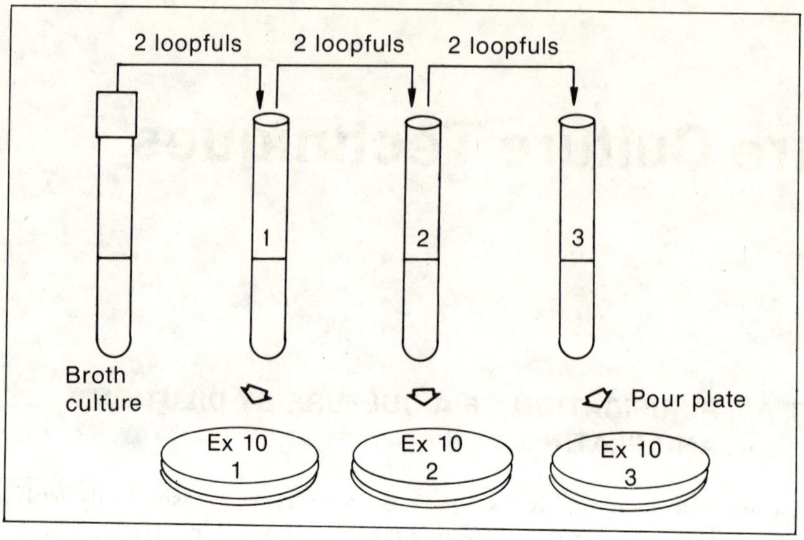

FIGURE 10–1

to the medium, or creating anaerobic conditions are some of the other techniques commonly used to isolate specific kinds of bacteria. However, effective use of these methods for isolating a pure culture requires considerable understanding of the physiological and biochemical characteristics of the organism. This generally precludes their use in a beginning bacteriology course.

The most widely used method of obtaining a pure culture involves dilution followed by pour plates (see Fig. 10-1) and, later, by streak plates (Fig. 10-3).

MATERIALS

1. Nutrient broth culture prepared in Experiment 8.
2. 5 tubes of nutrient agar (approximately 20 ml/tube).
3. 5 sterile Petri dishes.

PROCEDURE

PERIOD I

1. Label 3 sterile Petri dishes as dilution pour plates 1, 2, and 3 (Fig. 10-1) and 2 dishes as streak plates 1 and 2.

2. Obtain 5 melted nutrient agar tubes from the water bath (50C). Wipe the tubes dry and label 3 of them as 1, 2, and 3.

3. Flame and pour the remaining 2 tubes of agar into dishes labeled streak plates. Allow to solidify.

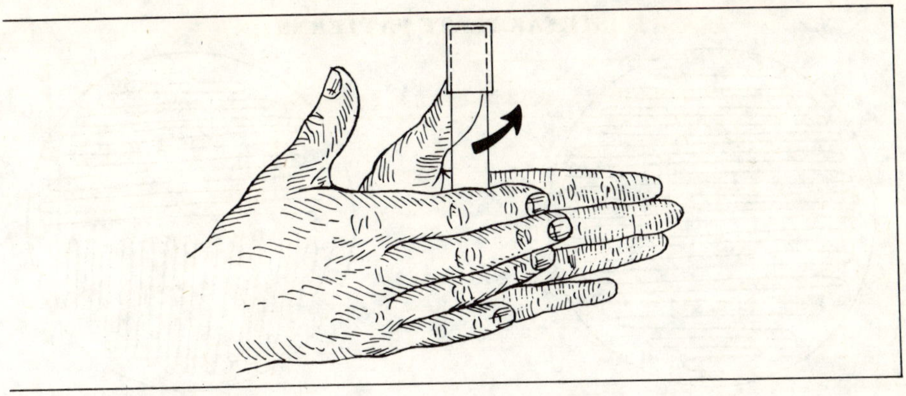

FIGURE 10–2. Proper mixing of inoculated agar tube for pour plate.

4. Mix the broth culture (prepared in Experiment 8) until it appears uniformly turbid.

5. a. Flame sterilize a loop and transfer 2 loopfuls of the broth culture to tube 1 (Fig. 10-1). Mix thoroughly by rotating the tube between the palms (Fig. 10-2).

   b. With a sterile loop, transfer 2 loopfuls of the culture from tube 1 to tube 2. Again, mix tube 2 thoroughly as in Step 5a.

   c. With a sterile loop, transfer 2 loopfuls of culture from tube 2 to tube 3. Mix tube 3 as above.

6. Pour the agar from tubes 1, 2, and 3 into the dishes labeled dilution plates 1, 2, and 3, respectively. Allow to solidify.

7. Prepare 2 streak plates by dipping the flame-sterilized loop into the broth culture (instructor will demonstrate the technique). Any two streak patterns shown in Figure 10-3 may be used. (Note: all patterns are designed to give well-separated colonies.)

8. Invert all the plates and incubate in your desk cabinet for 2 to 5 days.

PERIOD II

PART B. EXAMINATION OF COLONIES AND PREPARATION OF A STOCK CULTURE

Proof of purity of a culture can be ascertained by two general methods. (1) *Microscopic examination of stained smears prepared from the colony:* If the colony is the outgrowth of a single cell, the smear should show a single morphologic type. (A note of caution: even in a pure culture there is unavoidable variation such as in size of cells; the cells immediately after division are usually smaller than those cells

STREAK PLATE PATTERNS

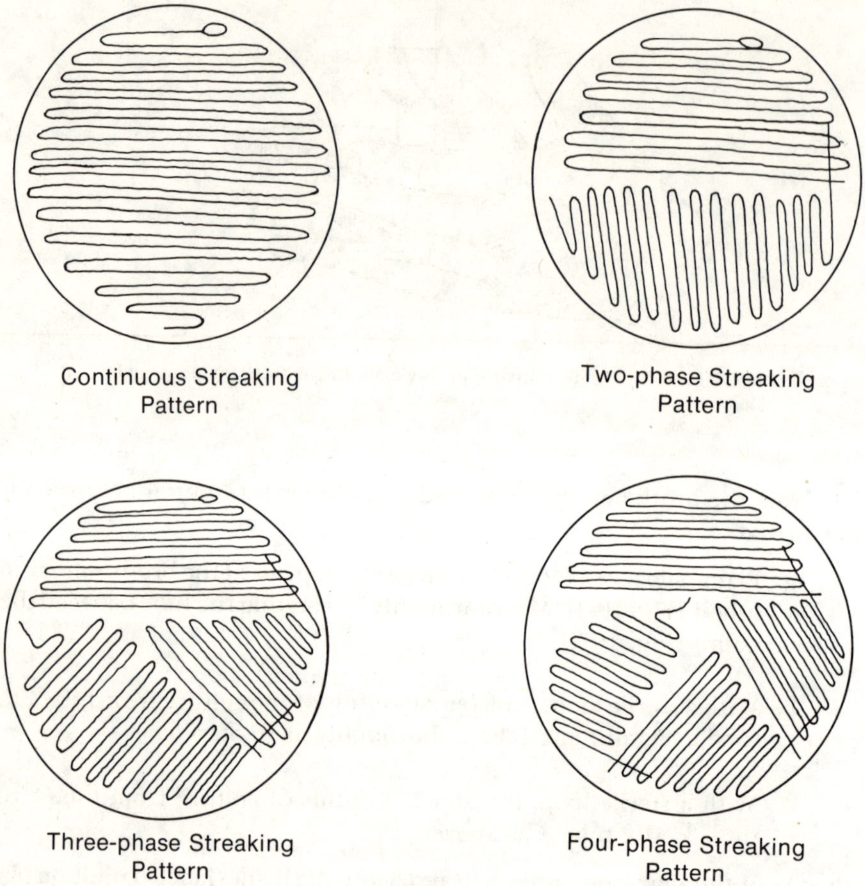

FIGURE 10–3

Choosing a streaking pattern is a matter of individual preference and judgment as to the population density of the sample. The figures above illustrate four of the patterns most commonly used. The object of these patterns is the continuous dilution of the inoculum to give well-separated colonies on the plates. For multiphase streak plates, students *must* flame the loop before starting each phase to remove all the cells which adhere to it. *Note slight overlap* into the previous phase on the first stroke to pick up a small inoculum.

about to divide.) (2) *Examination of colony morphology on both dilution and streak plates:* If the colonies on the plates are descendent (developed) from a pure culture, all of the colonies should be the same morphologic type within the usual limits of variation of individual colonies.

Isolated pure cultures must be maintained in *viable* (able to reproduce) cell condition for further study; to do this, one establishes a *stock culture,* the best possible representative subculture of the original culture. Viable stock cultures may be maintained on agar slants, usually of the medium most conducive to growth of that organism, and the viability of the stock culture may be lengthened by keeping it under refrigeration (4 to 9C). Another method of storing viable stock cultures is through lyophilization or freeze-drying. The lyophilized stock cultures may be kept for long periods of time (months or years) with only a gradual decrease in the viability of the original population.

MATERIALS

1. Dilution plates from first period of Experiment 10.
2. 2 tubes of melted nutrient agar medium.
3. 1 tube of nutrient broth.

PROCEDURE

1. Obtain 2 melted nutrient agar tubes and allow them to solidify on a slope so as to form slants. The instructor will show you an example of a well-made slant.

2. Examine the colonies on the dilution plates with the unaided eye. Look for uniformity in size of surface and subsurface colonies, colony color, pigments, and elevation. Record your observations in the space provided. To describe the colonies with the proper terms, refer to Figure 9-5.

3. The margin or edge of the colonies may be observed with the low power objective. Move the mechanical stage toward you to make a space for the Petri plate. Place the plate on the stage in an inverted position. Focus upon the colony with the low power objective. Record your observations in the space provided.

4. Observe streaked plates. Are there any well-separated individual colonies? Observe the colonies as in Steps 1 and 2 above.

5. If your observation of colonies indicates pureness of the culture, select one colony that is well isolated from the others yet representative of them. Make a circle around the colony with a marking pencil.

6. Using aseptic technique, prepare a subculture of the colony you selected by touching the center of the colony with a sterile wire needle, then transferring the growth on the tip of the needle to the nutrient agar slant. Streak the growth on the center of the slope, starting from the bottom and moving toward the top (as demonstrated by your instructor). Prepare another slant from the same colony. This culture will be used throughout the remainder of the course for the purpose of identification and classification; consequently you are expected to maintain a viable stock culture at all times. Likewise, it is imperative that you submit the complete study of your unknown with the specific name of the genus and, if possible, the name of the species.

7. Similarly inoculate 1 tube of nutrient broth. Incubate this tube as well as the 2 slants (stock cultures) in your desk cabinet for 2 to 5 days.

PERIOD III

1. Observe the growth of your culture on the agar slant. Record the form of growth using examples given in Figure 9-5.

2. Prepare a Gram stained smear by removing a small amount of growth from the agar slant. Record your results in the space provided. Note: culture age should not exceed 24 hours.

3. Keep one stock culture in your locker as working stock, and store the other in the refrigerator as permanent stock. Working stock should be transferred to a new slant every few weeks to maintain viability. Likewise, the permanent stock should be renewed sometime during the eighth week of the course. Remember, it is your responsibility to maintain a stock culture that is pure as well as viable.

RESULTS

Describe and draw morphologic and cultural characteristics of a purified unknown isolate in the table below.

Source of unknown:

Morphological and Cultural Characteristics	Word Description	Drawing
Cell shape and Gram reaction		
Pigment on NA		
Colony morphology		
Growth on agar slant		
Surface growth in nutrient broth		

DISCUSSION

1. In general, the viability period of a culture is lengthened when the culture is stored at refrigeration temperature rather than at room temperature. Why?

2. What is the reason for making a Gram stained smear with cultures incubated for 24 hours or less? If you are uncertain, you may refer to Experiment 4.

EXPERIMENT 11

Characterization and Differentiation of Bacteria

The gross and macroscopic appearances of bacterial cultures on different kinds of media are often useful in identification of the organisms. For example, colonies on an agar plate frequently demonstrate size, shape, color, margin, and elevation typical of certain species. These growth characteristics provide clues to the identity of the organisms and should always be included as a part of classifying them.

Cultural studies are usually made with nutrient broth, agar slants, and plates. Frequently, nutrient gelatin deeps and fluid thioglycollate medium are also used. In this experiment we will include the potato slant, although it is no longer used as a medium for systematic study of unknown microorganisms, because so many microbiologists before 1900 employed it in studying the cultural characteristics of their isolates. The potato slants are prepared from small slices of fresh potato and subsequently are sterilized in an autoclave at 121C for 5 minutes. Glucose-yeast extract–soft agar deeps are also included to study the motility of the unknown organisms.

The various ways in which bacterial motility may be demonstrated were discussed in Experiment 5. In this experiment only the motility agar will be used, but students may find it helpful to review the other procedures.

To describe cultural characteristics, one must use terms which are familiar to all microbiologists. Such terms are used in your textbook and in H. J. Conn and H. W. Jennison's *Manual of Microbiological Methods* (McGraw-Hill, 1957), V. B. Skerman's *Guide to the Identification of the Genera of Bacteria* (2nd ed., Williams and Wilkins, 1967), *Bergey's Manual of Determinative Bacteriology* (7th and 8th eds., Williams and Wilkins, 1957, 1974). The latter two books will be used by each student in identifying his unknown.

MATERIALS

1. 1 slant of nutient agar.
2. 6 nutrient gelatin deeps (see Appendix).
3. 5 tubes of fluid thioglycollate medium (see Appendix).

4. 5 motility agars (glucose–peptone–yeast extract–soft agar deeps) (see Appendix).

5. 5 slants of potato medium.

6. 1 tube each of pure cultures of *Proteus vulgaris* (1); *Staphylococcus aureus* (2); *Pseudomonas fluorescens* (3); *Enterobacter aerogenes* (4); your unknown culture (5).

PROCEDURE

PERIOD I

1. Make a fresh working stock culture of your isolate from Experiment 10 on a nutrient agar slant. Incubate the fresh slant in your desk. (Remember it is your responsibility to maintain a viable stock culture.)

2. **Gelatin stab.** Flame sterilize a wire needle, obtain a small amount of growth from your unknown, and aseptically make a deep stab inoculation (see Fig. 11-1). The wire needle with growth should be stabbed *straight* into the bottom of the tube. Remember, the gelatin deep should be inoculated while it is solid. Gelatin may become semi-liquid at warmer temperatures; if your gelatin deep is already semi-liquid, place it in an ice bath before inoculation to promote firmer gelling. Also make gelatin stabs with the pure cultures provided. Keep an uninoculated tube as control.

3. Inoculate your unknown in fluid thioglycollate medium as you inoculated it into the gelatin, except use a wire loop and mix the inoculum by rotating it between your palms. Do not shake (to avoid introduction of air) as this medium is designed to determine the oxygen requirement of your unknown. Do the same with the pure cultures provided.

4. Inoculate your unknown into the motility agar as in Step 2. Make as straight a stab as possible. Do the same with the pure cultures provided.

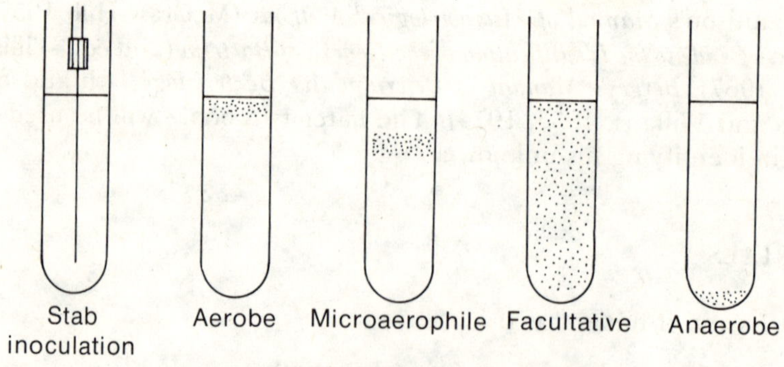

FIGURE 11–1. Proper stab inoculation of solid and semi-solid medium and location of microbial growth in fluid thioglycollate medium.

5. Inoculate the potato slants with the cultures provided as well as with your unknown. (Steam sterilized potato is soft—care must be taken when streaking; appearance of mashed potato is discouraged!)

6. Incubate all media in your desk until next period.

PERIOD II

1. **Gelatin liquefaction.** Gently place all 6 tubes of gelatin deep into the ice bath provided and keep them in the bath until the uninoculated control tube has gelled firmly. Observe the gelatin stabs and look at the types of growth and the presence or absence of liquefaction (loss of the ability to gel, even in the cold). It should be remembered that not all organisms grow in the gelatin deeps, but if growth does occur, liquefaction of gelatin may also occur. The ability of organisms to break down gelatin will be discussed in more detail in Experiment 14. If your culture did grow, or there was growth and liquefaction, you should refer to Figure 9-5 and describe your observations in the space provided.

2. **Effects of oxygen tension.** Observe the growth in fluid thioglycollate medium. (This medium has been used for the cultivation of anaerobes from body fluids and other sources because it has a low oxidation-reduction potential, i.e., the small concentration of agar present prevents reaeration of the medium by convection currents after sterilization. See *Difco Manual,* 9th ed., Difco Laboratories, Detroit, Michigan.) If your unknown grew only at the surface of the broth, your culture is considered *aerobic:* it requires oxygen for growth. However, if growth occurs throughout the tube, (from the surface to the bottom) your unknown culture is *facultative:* it is able to grow in the presence or absence of dissolved oxygen. As a rule, however, facultative organisms grow better in the presence of free oxygen (O_2 acting as the final hydrogen acceptor) than in the absence of free oxygen or in the presence of reducible organic compounds (either metabolically produced or added for the purpose). If the culture grows only at the bottom of the medium, the culture is said to be *anaerobic.* A few species are *microaerophilic:* they require less than atmospheric concentrations of oxygen for growth.

3. **Motility agar.** Observe all 5 tubes inoculated. An example of nonmotile growth is *S. aureus,* whose line of growth is straight; motile growth is represented by *P. vulgaris,* whose growth is diffused throughout the deep. Are the remaining three cultures motile? Record your observations in the space provided.

4. **Potato slant.** Examine the growth on the potato slant. Record your observations in terms of pigment production, form of growth, color of streak, etc.

RESULTS

1. Draw the pattern of growth that developed in the nutrient gelatin deep. If your culture liquefied the deep, refer to Figure 9-5 for proper terms of description.

2. Describe the type of growth in the fluid thioglycollate medium.

3. Sketch the type of growth in the motility agar deep.

NOTE: At the end of this experiment you will have obtained the following information about your unknown culture:

1. Cellular morphology: rod, coccus, spirillum, or filamentous.
2. Endospore formation (prepared on your own time).
3. Staining reaction: gram-positive or gram-negative, etc.
4. Cultural characteristics:
    a. Surface colony on nutrient agar.
    b. Submerged colony on nutrient agar.
    c. Motility: based on wet mount and motility agar.

d. Pigment production on nutrient agar and potato slant.

e. Growth characteristics in nutrient gelatin stab; gelatin liquefaction.

f. Effect of oxygen tension on growth (fluid thioglycollate medium).

GROWTH CHARACTERISTICS OF THE ORGANISMS

	Medium			
Culture	*Nutrient gelatin deep*	*Fluid thioglycollate medium*	*Motility deep*	*Potato slant*
Your unknown culture	Only surface growth no liquid	Fac	aerobic motile	fantastic growth
P. vulgaris	no growth	Fac	Non motile	growth
Bacillus ~~*S. aureus*~~ subtilis	Facultative	Fac	Non motile	growth
P. fluorescens	liquefaction stratiform	Aerobic	motile on top aerobic	growth
E. aerogenes	[illegible]	Fac	motile on top aerobic	growth

DISCUSSION

1. Define aerobic, facultative, microaerophilic organisms in relation to fluid thioglycollate medium.

2. A gelatin tube after standing in an ice bath for a period of time fails to gel. What is the reason for this?

3. What other methods can be used to determine motility of an organism?

EXPERIMENT 12

Viruses: Enumeration and Specificity of Bacteriophage

Viruses are obligate, intracellular parasites, submicroscopic and filtrable. All are composed of a nucleic acid core surrounded by some type of protein outer covering called the *coat* or *capsid**. There are many types of viruses: plant viruses, insect viruses, animal viruses, tumor-causing viruses, and they differ from one another with regard to transmission, infectivity, and replication. In this experiment, we will work with bacteriophages, viruses which attack bacteria, because of the ease with which they may be handled and because they are not believed to be pathogenic for man.

Bacteriophages come in many sizes and shapes, as shown in Table 12-1, and may contain either DNA or RNA, but never both. The T-even phages are the most complex (see Fig. 12-1). They have a polyhedral "head" measuring about $65 \times 95\mu m$ and a rather complicated contractile "tail" measuring $17 \times 115\ \mu m$. Further details on viral morphology and replication will be found in the text.

Replication of bacteriophages occurs only within living bacteria. They are very specific in their activity; a bacteriophage for one strain of bacteria within a species may not attack another strain of the same species. This relationship is called bacteriophage *host specificity*. If bacteriophage activity is limited to one kind of bacterium, it is often called *monovalent phage* and a certain virus capable of attacking more than one kind of bacterium for replication is called a *multivalent phage*.

The process by which phages infect the cell (by injection of phage nucleic acid), and the subsequent replication of more phage particles and destruction of the cell with release of virions, is called the *lytic cycle*. However, some phages (called *temperate* phages) may bring about a *lysogenic cycle* rather than a lytic cycle. This happens when the phage nucleic acid attaches to the bacterial chromosome and is replicated in the *prophage* state along with the bacterial chromosome. The culture is then said to be lysogenic; no phage particles are formed and the bacteria are not lysed. Each daughter cell, however, receives a copy of the prophage and thus carries the potential to produce phage under certain conditions. See text for further details of induction, transduction, and related genetic phenomena.

The lytic cycle (Fig. 12-2) begins with the attachment of tail fibers of a *virulent* phage to the cell envelope, followed by the fastening of tail pins and the base plate to a specific *receptor site* on a host cell. The cell surface has many receptor sites and is capable of adsorbing numerous phage particles simultaneously, but

*An infectious virus without a capsid (naked virus) is termed a *viroid*.

TABLE 12—1.
REPRESENTATIVE MORPHOLOGIC TYPES OF BACTERIOPHAGE

Examples (coliphages)	Basic morphology	Description	Nucleic acid type
T-even (T_2, T_4, T_6)		Angular head, contractile tail	double stranded DNA
T1 T5		Angular head, long non-contractile tail	double stranded DNA
T3 T7		Angular head, short non-contractile tail	double stranded DNA
ϕX174		Angular virion, no tail	single stranded DNA
fd M13 f1	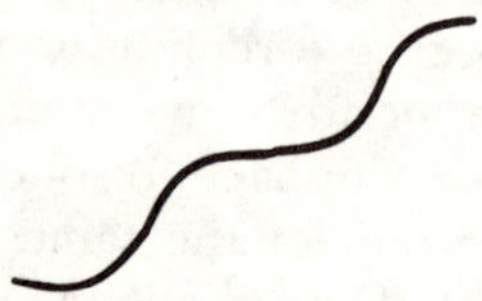	Flexible filament	single stranded DNA
f2 R17 fr		Angular virion, no tail	single stranded RNA

(From Lechevalier, H., and Pramer, D. *The Microbes.* Philadelphia: J. B. Lippincott Co., 1970.)

usually only one phage (of the same kind) actually injects its nucleic acid. The injected phage nucleic acid in the cell is now said to be in the *vegetative state* and the empty phage coat left outside the cell is called a *ghost.* The *vegetative phage DNA,* through a series of events, then takes control of the cell and initiates and directs the synthesis of more phage protein coat and nucleic acid. When these two parts are assembled, the maturation of phage is completed and new infectious phage particles *(virions)* are liberated from the host cell as the cell lyses. The number of particles liberated per host cell is called the *burst size.* The period from the es-

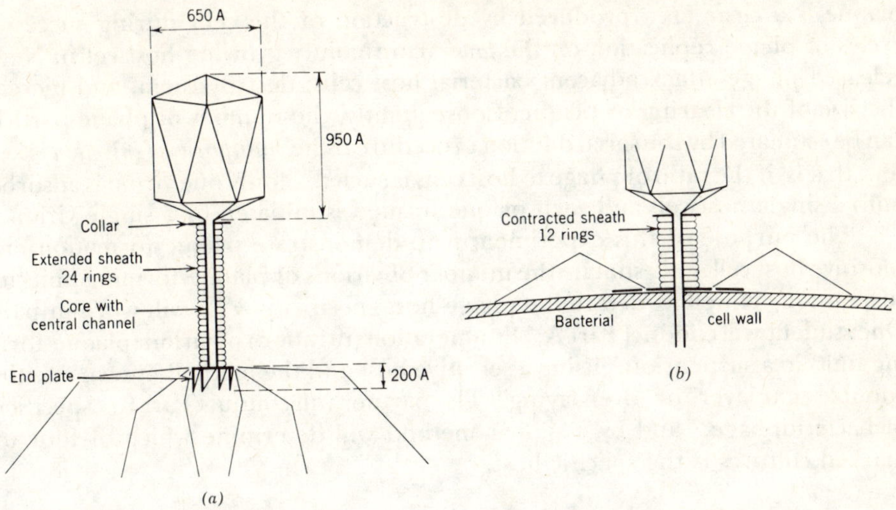

FIGURE 12–1. Diagrammatic representation of a T-even coliphage. (Modified from Luria, S. E., and Darnell, J. E., Jr.: *General Virology,* 2nd ed. New York: Wiley, 1967.)

tablishment of vegetative phage state to just before the liberation of new virions is called the *latent period.* If the host cell adsorbs too many phage particles, the host cell wall may be weakened and lysis may result without any phage replication. This phenomenon is often called "lysis from without."

If the lytic cycle occurs in a broth culture, the turbidity of the culture decreases; if the lytic process occurs on an agar plate culture, *plaques* are formed. The

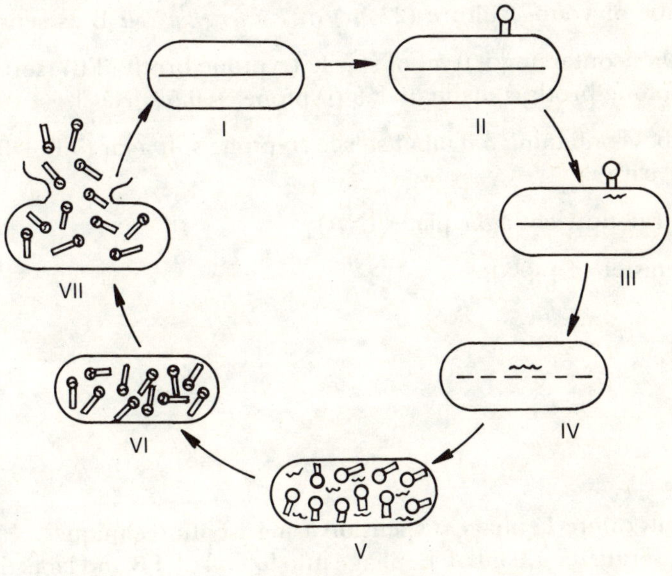

FIGURE 12–2

I. Uninfected host cell.
II. Phage attached at a specific receptor site.
III. Injection of phage nucleic acid. Phage DNA at this stage is called vegetative phage; empty phage coat (ghost) remains outside cell envelope.
IV. Phage DNA takes over genetic control of cell.
V. Phage nucleic acid and protein synthesized.
VI. Assembling of protein coat and nucleic acid and maturation of the phage.
VII. Release of phage particle (virion) upon host cell lysis.

plaque is a clear area produced by destruction of the cells during successive cycles of phage replication on the *lawn* (confluently growing host cells). Newly released phage attack adjacent bacterial host cells, destroy them, and increase the size of the clearing or plaque. Consequently, the number of phage particles can be estimated by the serial dilution procedure if the *multiplicity of infection* is kept low, that is, if the ratio of phage to host cells is such that only one virion is adsorbed onto a single host cell and each plaque formed is initiated by a single virion.

The purpose of this experiment is to demonstrate plaque formation on a sensitive host cell and estimate the number of virions or plaque-forming units in a suspension, as well as determine phage-host specificity. You will work in pairs. One student will conduct Part A, "Enumeration (titration) of virion (plaque-forming unit) in a suspension," using a sensitive host and the method known as either "double agar layer" or "over-laying". The partner will conduct Part B, "Specificity of bacteriophage," and by a similar method will determine which of four unmarked cultures is the specific host.

PART A. TITRATION OF PLAQUE-FORMING UNITS IN SUSPENSION

MATERIALS

1. 1 tube of T_3 bacteriophage suspension.

2. 1 tube of young culture (24 hr) of *Escherichia coli* B as sensitive host.

3. 2 tubes containing 9.0 ml of sterile tryptone broth (TB) (serial dilution blanks). Tryptone broth is made of 1% tryptone + 0.5% NaCl.

4. 4 tubes containing 3.0 ml of sterile tryptone soft agar (TB + 0.7% agar) (in the water bath).

5. 4 sterile nutrient agar plates (NA).

6. A canister of pipettes.

PROCEDURE

PERIOD I

1. Serially dilute T_3 phage suspension using aseptic techniques.
   a. Transfer 1.0 ml of T_3 phage into 9.0 ml of TB and label the tube as 10^{-1}. Mix thoroughly by rotating the tube between your palms.
   b. Transfer 1.0 ml of T_3 phage suspension from the 10^{-1} tube into a second TB tube. Mix well and label the tube as 10^{-2}.

2. Obtain 4 tryptone soft agar tubes from the water bath set at 45 to 50C. Pipette 0.2 ml of *E. coli* B into each of 4 tryptone soft agar tubes. Mix thoroughly, label all 4 tubes Ec, and return them to the water bath.

3. Arrange nutrient agar plates in a row in front of you and label 1 plate

as 10^{-1}, the second plate as 10^{-2}, the third plate as 10^{-3}, and the fourth plate as the lawn control.

4. a. Remove 1 inoculated tube of tryptone soft agar (Ec) from the water bath. Wipe off all of the water from the surface of the tube.

   b. Pipette aseptically 0.1 ml of undiluted T_3 phage suspension into the soft agar.

   c. Mix the contents thoroughly by rotating the tube between your palms.

   d. Pour its contents aseptically onto the layer of nutrient agar in the plate labeled 10^{-1}.

   e. Replace the Petri dish lid and distribute the contents evenly over the surface by carefully tilting the dish back and forth.

   f. Set the plate aside and allow to solidify.

5. Repeat the above process using the 10^{-1} and 10^{-2} T_3 phage dilutions you have prepared. Remember, the contents of 10^{-1} phage-culture preparation will be poured into the plate labeled 10^{-2}, since 0.1 ml of the 10^{-1} suspension was added. Likewise, the contents of 10^{-2} phage-culture preparation will be poured into the plate labeled 10^{-3}.

6. a. Remove the fourth tube (Ec) from the water bath (do not add phage at this time) and aseptically pour the entire contents onto the plate labeled control. Distribute the contents evenly over the surface of the plate as in Step 4.

   b. Allow to solidify.

   c. Sector the bottom of the control plate into two equal parts and on one half make 3 small circles as shown by your instructor; label circles with 1, 10^{-1} and 10^{-2}.

   d. Aseptically transfer a loopful of undiluted T_3 phage suspension to circle 1 and similarly a loopful of phage suspension 10^{-1} to circle 10^{-1}; do the same with 10^{-2} phage dilution.

   e. Allow the phage preparation to diffuse into the lawn. This process is often called a "spot test" and is very useful in selecting susceptible host cultures from related species. The uninoculated half of the plate will serve as a control. Confluent growth of host cells, i.e., a lawn, is necessary for formation of clear plaques.

7. After the "over-layered" agar used in Step 5 has solidified, invert the plates and incubate for 1 to 2 days in the 37C incubator.

PERIOD II

1. Examine each prepared phage assay plate under a colony counter. Small, clear spots distributed in the confluent lawn of the sensitive bacteria are plaques (see introduction).

2. Count the number of plaques on each dilution plate. If the plaque number is too great, sector the plate into 4 equal parts, count 1 sector, and multiply by 4. This will give the approximate number of plaques on the plate. Record the number of plaques formed on each plate in the appropriate place. Calculate the approximate number of virions per ml of the original T_3 phage suspension by multiplying the number of plaques by the appropriate dilution, i.e., if 20 plaques are found in the 10^{-3} dish, then 20×10^3 virions/ml were in the original T_3 phage preparation.

3. Examine the control plate. Note the complete clearing of spot 1. Such clearing is often called *confluent lysis* and is caused by too many virions placed in a relatively small area of the lawn. Observe the successive reduction in lysis of the lawn as dilution increases. Record your observations in the space provided.

PART B. DETERMINATION OF A SPECIFIC HOST FOR T_3 PHAGE FROM FOUR CODED CULTURES (PARTNER'S ASSIGNMENT)

MATERIALS

1. 1 tube prediluted T_3 bacteriophage.

2. 1 tube each of 24 hr cultures of *Escherichia coli* B, *E. coli* K, *Staphylococcus aureus,* and *Pseudomonas fluorescens.* (These names will not appear on the tubes, only coded numbers. Only your instructor knows the identity of the cultures.)

3. 4 tubes of melted tryptone soft agar. (See Part A Materials section for the composition.)

4. 4 sterile nutrient agar plates.

5. A canister of pipettes.

PROCEDURE

PERIOD I

1. For each unknown culture, label one nutrient agar plate, using the coded numbers.

2. Label 4 melted tryptone soft agar tubes in the same fashion.

3. Aseptically pipette 0.2 ml of a coded culture and transfer it into a correspondingly marked tube of tryptone soft agar. Do the same with the remaining 3 cultures, and return all 4 inoculated soft agar tubes to the water bath (45 to 50C).

4. a. Take 1 tube of tryptone soft agar inoculated in Step 3 from the water bath. Wipe off all the water from the outside of the tube and

pipette aseptically 0.1 ml of T_3 phage into the soft agar. Mix thoroughly by rotating the tube between the palms of your hands.

b. After mixing, aseptically pour entire contents onto the surface of the nutrient agar plate labeled with the corresponding code number. Replace the lid of the plate immediately to avoid contamination from the air.

c. Distribute the poured agar contents evenly throughout the plate by carefully tilting the plate back and forth. Allow to solidify.

5. Do the same for the remaining 3 inoculated soft agar tubes in the water bath.

6. Allow to solidify, invert the plates, and incubate them in the 37C incubator for 1 to 2 days.

PERIOD II

1. Under a colony counter, examine each Petri plate culture for plaque formation. Remember, T_3 phage is specific for *E. coli* B and the other 3 cultures used will not serve as host for the phage. Make a decision as to which of the 4 cultures was *E. coli* B.

2. Count the number of plaques formed on the sensitive host and determine the number of plaque-forming units in the original prediluted T_3 phage suspension. Number of plaques $\times 10^1$ would give you T_3 virion/ml.

3. The instructor will announce the identity of codes at this time. Have you arrived at the correct decision? Record your observations in the space provided.

RESULTS

PART A. TITRATION OF PLAQUE-FORMING UNITS IN SUSPENSION

1. a. Record the results of phage titration.
   Number of plaques formed on 10^{-1} plate ____________________
   Number of plaques formed on 10^{-2} plate ____________________
   Number of plaques formed on 10^{-3} plate ____________________
   Therefore, the original T_3 phage suspension contained__________ virions or plaque-forming units per ml.

   b. Describe the results of the spot test on your control plate in terms of degree of lawn lysis with increased dilution.

PART B. DETERMINATION OF A SPECIFIC HOST FOR T_3 PHASE FROM FOUR CODED CULTURES

1. a. Record your observations on plaque formation on the 4 coded cultures.
   Code___Organism_______________ Plaques__________________
   Code___Organism_______________ Plaques__________________
   Code___Organism_______________ Plaques__________________
   Code___Organism_______________ Plaques__________________

   b. Record the number of T_3 virions in the prediluted suspension: _______________virions or plaque-forming units/ml.

DISCUSSION

1. What is the significance of "phage-typing" as a tool in the identification and classification of the unknown organism?

2. Would a large phage tend to produce large or small plaques? Why?

3. What limits the final size of a plaque?

4. Does the 30 to 300 rule apply to counting plaques as well as colonies on a plate?

SECTION

II

GROWTH AND METABOLISM OF BACTERIA

EXPERIMENT 13

Minimal Growth Requirements for Bacteria

Bacteria, like all other living things, have certain nutritional requirements. Although we often do not think of water as a nutrient, it is required by all microorganisms. The preservation of many foods (honey, jellies, preserves, salted meat, etc.) and other products is based on the principle that there is insufficient available water for growth of most types of microorganisms. Water is considered a universal solvent, provides elements H and O, and is an essential ingredient of protoplasm.

Chemical analysis of bacterial cells has shown that carbon and nitrogen constitute the bulk of cell dry weight. Microorganisms vary, however, in their ability to utilize different forms of carbon and nitrogen present in the environment. Some bacteria can utilize CO_2 as their sole carbon source *(autotrophs)*, while others require complex organic substrates *(heterotrophs)*. Similarly, some microorganisms can obtain their nitrogen from simple compounds such as ammonia or nitrates or even from nitrogen gas (through nitrogen fixation, see Experiment 26). Others require more complex nitrogenous compounds in addition to growth factors and vitamins (such as would be found in yeast extract, serum, etc.).

Nutrients such as S, P, and K are required by bacteria, but most can be assimilated as simple inorganic salts: $MgSO_4$, KCl, or $(NH_4)_2HPO_4$. Sulfur is assimilated as a constituent of sulfhydryl-containing amino acids (R–SH), and phosphorus is used in synthesis of nucleic acids (DNA, RNA) and various enzymes (NADP, NAD, FAD, etc.). Potassium is one of the principal intracellular cations and is a cofactor for certain enzymes. Other elements such as calcium, iron, magnesium, manganese, and even zinc are needed in small amounts to act as cofactors for enzymes. Still other elements, referred to as trace elements (copper, molybdenum, etc.), are required in very small amounts for certain enzymes. Such minute quantities are needed that often the glassware in which the organisms are grown provides a sufficient supply.

Not only must microorganisms have all the essential elements needed for the synthesis of new cell material, but they must also be provided with an energy source. Those organisms capable of utilizing CO_2 as their sole C source are further separated on the basis of their energy source. *Photoautotrophs (photolithotrophs)* utilize the sun's light as their energy source, whereas *chemoautotrophs (chemolithotrophs)* are capable of obtaining their energy from oxidation of simple inorganic elements or compounds. Organisms requiring organic carbon are collectively known as *heterotrophs (chemoorganotrophs).* A small group of organisms requiring organic carbon and whose growth is stimulated in the presence of the sun's light are frequently referred to as *photoorganotrophs.* Students should become familiar with these terms and use them where applicable.

While the nutrition of bacteria cannot be discussed here in any depth, we will consider in the next two experiments how bacteria are able to utilize complex substrates for growth. In this experiment we will demonstrate that even the simplest of bacterial culture media must provide certain minimal nutrients in order to support the growth of nonfastidious bacteria. Furthermore, it will be seen that commonly encountered bacteria differ in their ability to grow on simple media. Other requirements, such as proper gaseous environment, moisture, pH, and temperature, are briefly discussed in Experiments 16, 24, and 29.

PERIOD I

MATERIALS

1. 1 agar deep (20 to 25 ml) containing 1.5% washed agar* in double distilled water, marked A.

2. 1 agar deep (20 to 25 ml) containing 1.5% washed agar in salts solution†, marked AS.

3. 1 agar deep (20 to 25 ml) containing 1.5% washed agar and 0.5% glucose in salts solution, marked AGS.

4. 1 agar deep (20 to 25 ml) containing 1.5% washed agar, 0.5% glucose, and 0.5% peptone in salts solution, marked AGPS.

5. 1 agar deep (20 to 25 ml) containing commercial brain-heart infusion agar, marked BHI.

6. Cultures of the following organisms on trypticase–soy agar slants: *Escherichia coli, Streptococcus lactis, Staphylococcus aureus, Pseudomonas aeruginosa* and *Neisseria catarrhalis.*

7. 5 sterile Petri dishes.

PROCEDURE

1. With a marking pencil, divide the bottom of each of 5 sterile Petri dishes into 6 equal sectors.

---

*Washed agar will be prepared prior to class. The procedure is to suspend either Noble Agar (Difco) or Ionagar (Oxoid) in an excess of cold double distilled water (large Erlenmeyer flask with magnetic stirring bar). The suspension is kept cold and the wash water changed 5 to 6 times over a period of 48 hr. The agar is dried on a Buchner funnel filter by successive washes with absolute alcohol and acetone. The agar is then air dried under gauze or in a dessicator jar under vacuum at 37C until suitable for use. Although some batches of Noble Agar or Ionagar may be satisfactory for this experiment without washing, the above procedure is highly recommended for consistent results. Washed agar prepared in this fashion is also excellent for immunologic work (immunodiffusion and immunoelectrophoresis).

†The salts solution is designed to provide several of the key elements necessary for growth and contains per 100 ml: 0.1 g $NH_4H_2PO_4$, 0.1 g K_2HPO_4, 0.02 g $MgSO_4$, and 0.5 g NaCl. The salts are dissolved in distilled water and the pH adjusted, if necessary, to 7.0 to 7.2.

2. Label each dish to correspond to one of the types of media being employed, i.e., A, AS, AGS, etc.

3. Pour the melted, cooled (45 to 50C) agar deep into the appropriately labeled dishes and allow the agar to solidify.

4. With a sterile loop, remove a *small* amount of growth from the TSA slant of *E. coli* and streak it on the corresponding sector of the "A" plate. Start near the center and streak toward the periphery. Do not dig into the agar. Flame loop.

5. Repeat this procedure for each of the other four types of media, flaming your loop between steps.

6. Repeat Steps 4 and 5 with each of the other 4 cultures.

7. Note that the sixth sector of each plate is uninoculated. It will serve as a sterility check for the medium and to show whether you are contaminating your plates with airborne microorganisms while you work.

8. Invert the plates and incubate at 37C for 2 days or 30C for 5 days.

PERIOD II

PROCEDURE

1. Examine the plates for growth and record in the Results section whether growth was absent (–), poor (+), moderate (+ +), or heavy (+ + +).

RESULTS

Bacterial Species	Media				
	Agar (A)	Agar + *Salts (AS)*	*Agar + Glu + Salts (AGS)*	*Agar + Glu + Peptone + Salts (AGPS)*	*BHI*
Escherichia coli	−	−	+	+	+
Streptococcus lactis	−	−	−	+	+
Staphylococcus aureus	−	−	−	+	+
Pseudomonas aeruginosa	−	+	+	+	+
Neisseria catarrhalis	−	−	−	+	+

6) − − + + +

DISCUSSION

1. What essential elements were provided in the salts solution and why is each of these required?

2. Can all bacteria obtain their nitrogen from simple ammonium compounds?

3. Where would you expect to find or be able to isolate the species of bacteria used in this experiment?

EXPERIMENT 14

Activities of Bacterial Exoenzymes

Morphological and cultural characteristics must be determined before unknown bacteria can be identified. However, these characteristics alone can rarely give sufficient information for identifying unknown bacteria as to genus and species.

If you are to identify a certain unknown pure culture, you must accumulate additional information by studying physiologic, pathogenic, and serologic properties, bacteriophage specificity, and other characteristics of the organism. Since a complete study of all these properties is beyond the scope of an introductory bacteriology course, this experiment will have to be limited to those physiologic characteristics which are easily demonstrable and which are most commonly used in the identification of organisms.

The determination of the types of hydrolytic exoenzymes (hydrolases) present is frequently used as an aid to bacterial identification. *Exoenzymes* (organic proteinaceous catalytic substances) are secreted from a cell and catalyze reactions outside the cell. The function of the exoenzymes is to break down (hydrolyze) large, complex, often water insoluble, food materials into simpler, soluble, and more immediately available foods.

The activities of enzymes, regardless of whether they are exo- or endoenzymes, are very specific: a certain enzyme can generally attack only a certain substrate; in fact, its activity is usually confined to one step among a series of reactions needed to break down a large molecule into smaller, useful foods. The ability to utilize complex foods such as polysaccharides, proteins, and fats is based on the organism's genetic capacities. For example, some organisms may be able to break down cellulose (a polysaccharide) through the production of the extracellular enzyme, cellulase; other organisms are incapable of hydrolyzing cellulose because they lack the gene coding for cellulase production. Consequently, the ability or inability to hydrolyze a specific substrate is an extremely useful property in identifying bacteria.

In this experiment you will demonstrate the presence or absence of three classic exoenzymes produced by certain bacteria.

1. Demonstration of Starch Hydrolysis (By Bacterial Amylase). Starch is a complex polysaccharide and unless it is broken down into smaller, more soluble compounds, the cells are unable to utilize it as a source of food. Production of amylase is easily demonstrated by growing the culture on nutrient media containing starch as the source of carbon or energy. As a rule, when starch is broken down by amylase, maltose (a soluble disaccharide) is the product.

Evidence of hydrolysis of starch to maltose or to glucose is shown by testing with Gram's iodine solution. The reagent produces a dark blue color when it comes in contact with starch, but not with maltose or glucose.

$$(C_6H_{12}O_6)_x \xrightarrow[H_2O]{\text{Bacterial amylase}} \text{Maltose} \xrightarrow[H_2O]{\text{Bacterial maltase}} \text{Glucose}$$

$$(C_6H_{12}O_6)_x + \text{Gram's iodine} \longrightarrow \text{blackish-blue color reaction}$$

$$\text{Maltose or glucose} + \text{Gram's iodine} \longrightarrow \text{no color}$$

2. Demonstration of Casein Hydrolysis. Casein is milk protein and the average cow's milk contains approximately 2.5 per cent of this substance by weight. A typical protein, casein is a complex polymer of amino acids and is insoluble in water. Certain bacteria can hydrolyze casein to peptides and amino acids and it is ultimately used by the cell as a source of energy as well as for synthesis of cellular components. Actually, the hydrolysis involves a series of enzymes and they are collectively called *caseinase*. Hydrolysis of casein by bacterial exoenzymes is demonstrated by the addition of 1 per cent HCl. Protein will be precipitated in an acidic environment, but products from hydrolyzed proteins such as proteoses, polypeptides, and amino acids will not be. A representative reaction is shown below:

$$\text{Casein} \xrightarrow[H_2O]{\text{Proteinase or protease}} \text{Proteoses} \xrightarrow[H_2O]{\text{Peptidase}} \text{Peptides and polypeptides} \xrightarrow[H_2O]{\text{Peptidase}} \text{Amino acids}$$

$$\text{Casein} \xrightarrow[H_2O]{\text{"Caseinase"}} \text{Hydrolyzed products} \xrightarrow{1\%\ HCl} \text{no ppt}$$

$$\text{Casein} \xrightarrow{1\%\ HCl} \text{white ppt}$$

3. Demonstration of Gelatin Hydrolysis. Gelatin, unlike casein, is not a typical protein. It dissolves in boiling water, gels when cooled to 20C or lower, and is not precipitated by lowering the pH, as are casein and other proteins. Gelatin is obtained from collagen, a protein found in skin, bones, and tendons, lacks tryptophan, and contains only small amounts of other important amino acids. Hydrolysis of gelatin can be demonstrated in two ways: (1) *Gelatin liquefaction* (as demonstrated in Experiment 11): when microorganisms hydrolyze gelatin, its solidifying or gelling ability is destroyed and this lack of gelling is evidence of gelatinase production. (2) *Precipitation by heavy metal:* native gelatin can be precipitated by acidified heavy metal solutions, but hydrolyzed products of gelatin will not be.

$$\text{Gelatin} \xrightarrow[H_2O]{\text{Gelatinase}} \text{Hydrolyzed products} \xrightarrow{HCl \cdot HgCl_2} \text{no ppt}$$

$$\text{Gelatin} \xrightarrow{HCl \cdot HgCl_2} \text{white ppt}$$

4. Demonstration of Fat (Lipid) Hydrolysis. Fats are glycerol esters of various fatty acids. For example, butterfats are chiefly composed of glycerol combined with butyric, oleic, and palmitic acids. When bacteria grow and hydrolyze fats, the products formed are fatty acids and glycerol; subsequently, both products are further metabolized as sources of energy and carbon. Evidence of fat hydrolysis *(lipolysis)* is obtained by testing for the presence of fatty acids which are freed from glycerides. Long-chain fatty acids, such as stearic and oleic, form soap when they react with salts of metals, such as Na, K, Ca, and Cu.

$$\begin{array}{l} CH_2\text{-}O\text{-}COC_{17}H_{35} \\ | \\ CH\text{-}O\text{-}COC_{17}H_{35} \\ | \\ CH_2\text{-}O\text{-}COC_{17}H_{35} \\ \text{(Tristearin [fat])} \end{array} \xrightarrow[H_2O]{\text{lipase}} \underset{\text{(Stearic acid)}}{3C_{17}H_{35}COOH} + \begin{array}{l} CH_2OH \\ | \\ CH\text{-}OH \\ | \\ CH_2\text{-}OH \\ \text{(Glycerol)} \end{array}$$

$$\text{Fat} \xrightarrow{CuSO_4} \text{no soap}$$

$$\text{Fatty acid} \xrightarrow{CuSO_4} \underset{\text{(blue ppt)}}{\text{copper soap}} + H_2SO_4$$

MATERIALS

1. 1 plate of nutrient agar.

2. 1 plate of starch agar (NA + 0.4% "soluble starch").

3. 1 plate of casein agar (sodium caseinate, 2.0 g; glucose, 1.0 g; $MgSO_4$, 0.2 g; K_2HPO_4, 0.2 g; and $FeSO_4$, trace, per 1000 ml distilled water).

4. 1 plate of gelatin agar (NA + 0.4% gelatin).

5. 1 plate of butterfat agar (NA + 1% butterfat). *(See note, Procedure, Period II, Step 5.)*

6. Gram's iodine.

7. 1% HCl solution.

8. $HCl \cdot HgCl_2$ ($HgCl_2$, 1.5 g; conc. HCl, 20 ml, diluted to 100 ml with distilled water).

9. Saturated aqueous solution of $CuSO_4$.

10. Pure cultures of *Pseudomonas fluorescens* (1), *Bacillus cereus* (2), *Enterobacter (Aerobacter) aerogenes* (3), and your unknown culture (4).

11. 3% H_2O_2.

PROCEDURE

PERIOD I

1. Invert the plates of nutrient, starch, casein, gelatin, and butterfat agar media. Sector each plate into 4 equal parts with a marking pencil. Label the sectors with numerals 1 through 4.

2. Using a sterile wire loop, inoculate sector 1 of each medium with culture 1. Spread the growth on the sector to about the size of a dime. Do the same for cultures 2, 3, and 4.

3. Incubate all the inoculated plates in an inverted position at 30C for 2 to 5 days.

PERIOD II

After an incubation period of 2 to 5 days, observe the growth of each culture on all of the media. Record the appearance of the growth in the space provided.

1. **Test for Catalase.** Catalase, one of the respiratory enzymes which split hydrogen peroxide into water and oxygen, may be detected by placing one drop of 3 per cent H_2O_2 over the growth on the nutrient agar plate. If bubbles of oxygen are liberated, the organisms have produced catalase.

$$H_2O_2 \xrightarrow{\text{catalase}} H_2O + \tfrac{1}{2}O_2$$

Most aerobic and facultative microorganisms are able to produce this enzyme. Record your results in the space provided.

2. **Test for Starch Hydrolysis.** Flood the starch agar plate with Gram's iodine solution. Allow the reagent to react for 2 minutes. The iodine reacts with starch to form a blue compound, but if the starch has been hydrolyzed to maltose or glucose by the action of bacterial *amylase,* no blue color will be formed. Usually it is the medium at the periphery of the growth which becomes clear. The reading should be made within 3 minutes as the starch-iodine complex will fade rapidly. If you have any questions regarding your results, consult your instructor. Record your observations in the space provided.

3. **Test for Casein Hydrolysis.** As stated in the introduction, most proteins can be precipitated by the addition of 1 per cent HCl. Flood the casein agar with the reagent and allow the reagent to react for about 5 minutes. Pour off the excess reagent into the bench sink before you make your observation. Observe the dish against a lighted area. If the casein has been hydrolyzed to proteoses, peptides or beyond by the action of caseinase, a clear zone will appear at the periphery of the culture. The unhydrolyzed area should be opaque from the precipitated proteins. Record your observations in the space provided.

4. **Test for Gelatin Hydrolysis.** The procedure for testing whether or not gelatin hydrolysis has taken place is similar to the test for casein hydrolysis. Flood the gelatin medium with $HCl \cdot HgCl_2$ reagent for 5 minutes. Pour off the excess reagent, read the results and record. The clear zone at the periphery of the growth indicates gelatinase production and a whitish, opaque appearance indicates gelatinase was not produced by the culture.

5. **Test for Fat Hydrolysis (Lipolysis).** Flood the butterfat agar with a saturated solution of aqueous $CuSO_4$. Allow the reagent to react with the medium for 15 minutes. Pour off the excess reagent and let the plate stand for about 10

minutes before reading the results. Examine fat globules *under, above,* and *at the periphery* of the growth. If lipase has been produced by the culture, fat globules will have become opaque and greenish-blue in color, indicating lipolysis of the fat. If the globules are unhydrolyzed, they will show no marked color reaction. Record all of your observations in the space provided.

NOTE: One of the drawbacks in determining lipase production is the difficulty of bringing the lipid substrates into intimate contact with the organisms. This contact is essential if the lipase is to be well diffused into solid culture medium. Hence, when you prepare the medium, make certain that lipids are well homogenized with the remaining nutrients in the medium.

At the end of this experiment you will have obtained the following information about your unknown in addition to that listed on page 97.

1. Catalase production
2. Starch hydrolysis
3. Casein hydrolysis
4. Gelatin hydrolysis
5. Butterfat hydrolysis (lipolysis)

Gel) 1) Pseudomonas Flourescens
2) Bacillus subtillis
3) Enterobacter aerogenes

Unknown

Circular pulvinate

EXPERIMENT 15

Fermentation of Carbohydrates

It is not uncommon for two different bacterial cultures to be very similar in their morphologic and cultural characteristics, yet show striking differences in their ability to utilize various carbohydrates and in the end products produced from carbohydrate metabolism (fermentation). For example, one culture may produce only a single acid *(homofermentation)* while another may produce many different acids or alcohols (*heterofermentation,* frequently referred to as *mixed acid fermentation*), with or without production of gases such as CO_2, CH_4, H_2S, and H_2. These characteristics are indispensable in classifying the organisms into certain taxonomic groups.

Carbohydrates are composed of either many units of monosaccharides of the same kind *(homopolymers)* or more than one kind of monosaccharide *(heteropolymers).* Polysaccharides may also be conjugated with other substances such as lipids *(lipopolysaccharides).* Starch, a homopolymer, is a carbohydrate composed of a long chain of glucose molecules. A disaccharide consists of two monosaccharide units and they may be two identical sugars (i.e., maltose, which is composed of two glucose molecules), or two entirely different sugars (i.e., sucrose, which is composed of glucose and fructose, or lactose—milk sugar—which is composed of glucose and galactose).

Carbohydrates, whether soluble, insoluble, di- or polysaccharides, in order to be used as food for energy or synthesis of cell constituents, must first be hydrolyzed by the microorganisms to monosaccharides (soluble). Monosaccharides such as glucose, galactose, and fructose may then be used as an energy source (dissimilation) and for the synthesis of cell material (assimilation) via complex metabolic pathways within the cells.

Carbohydrates are usually utilized by the two most common metabolic processes, *fermentation* and *respiration.* Fermentation is defined as the class of energy-yielding biological oxidation–reduction reactions in which organic compounds—metabolic intermediates—serve as the ultimate electron or hydrogen acceptors. Respiration differs from fermentation in that molecular oxygen is the ultimate electron or hydrogen acceptor. The organic electron acceptors in fermentation are usually reducible metabolic intermediates derived by oxidizing the sugar; consequently the amount of energy made available for growth and reproduction of the cell by fermentation is much less than that by respiration. Most of the available energy is still stored in end products. For example, the fermentation of glucose by *Lactobacillus arabinosus (homofermentation)* yields only lactic acid as the end product:

$$\underset{\textbf{(glucose)}}{C_6H_{12}O_6} \xrightarrow[H_2O]{\textit{L. arabinosus}} \underset{\textbf{(lactic acid)}}{2CH_3CHOH \cdot COOH}$$

A considerable amount of energy is still stored in lactic acid, which may be oxidized to CO_2 and H_2O by other organisms under suitable conditions. Note again that only one product, lactic acid, is formed in homofermentation.

An example of heterofermentation or mixed acid fermentation is the fermentation of glucose by *Enterobacter (Aerobacter) aerogenes.*

$$C_6H_{12}O_6 \xrightarrow[H_2O]{\textit{E. aerogenes}} CH_3CHOH \cdot COOH + CH_3 \cdot COOH$$
$$+ HCOOH + COOH \cdot CH_2CH_2 \cdot COOH$$
$$+ CH_3CH_2OH + CH_3 \cdot CHOH \cdot CHOH \cdot CH_3$$
$$+ CO_2 + H_2, \text{ etc.}$$

(glucose) **(lactic acid, acetic acid, formic acid, succinic acid, ethyl alcohol, 2,3-butylene glycol, carbon dioxide, hydrogen, etc.)**

Note that in the heterofermentation of glucose there is a considerable array of products formed in addition to gases. Again, much of the energy is tied up in end products.

Comparison of fermentation and respiration in terms of energy yield is best demonstrated by alcoholic fermentation and respiration of glucose by yeast cells.

Fermentation:

$$C_6H_{12}O_6 \xrightarrow[H_2O]{\text{yeast cells}} 2CH_3CH_2OH + 54\text{ kcal}$$

Respiration:

$$C_6H_{12}O_6 \xrightarrow[H_2O\,+\,O_2]{\text{yeast cells}} 6CO_2 + 6H_2O + 688\text{ kcal}$$

There are several methods of determining whether a microorganism can utilize carbohydrates as food for energy or synthesis of cellular materials. One of the oldest methods is the auxanographic procedure in which various dry carbohydrates are added to seeded basal medium containing no carbon source. Added carbohydrates diffuse into the medium and if the seeded culture is capable of utilizing them, the growth of the culture occurs around the diffused carbohydrate. The second method involves quantitative determination of added carbohydrate before and after the growth of the organism in a suitable medium. Reduction in the initial amount of carbohydrate after incubation indicates utilization of the carbohydrate by the organism. The third method is qualitative and quantitative measurement of the products of carbohydrate metabolism. However, this method requires a considerable knowledge of carbohydrate biochemistry. The fourth method is a routine qualitative test designed to permit

KNOWN: Glucose, Sucrose, Lactose Mannitol

TWO TYPES OF FERMENTATION TUBES

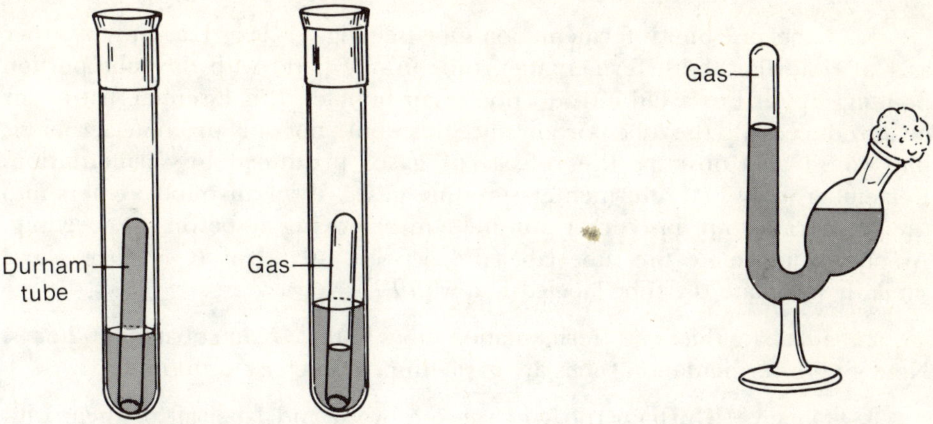

Durham fermentation tube - a small inverted vial to trap gas

Smith fermentation tube - gases produced are trapped in vertical tube

FIGURE 15–1

detection of carbohydrate utilization by noting a change in the pH of the culture medium. The medium generally consists of nutrient broth plus a carbohydrate and suitable indicator such as brom-cresol-purple (BCP). If the microorganisms ferment the carbohydrate and produce acid(s), the indicator (BCP) will change color from purple (pH 7) to yellow (pH 5.4), and gases, if produced, will be trapped in the Durham tube (Fig. 15-1). Thus the evidence of carbohydrate metabolism is largely circumstantial rather than direct, as in the second method. This experiment uses the fourth method to demonstrate various types of fermentation.

PERIOD I

MATERIALS (work with your partner)

1. 2 Smith fermentation tubes with tryptone (1.0%) yeast extract (0.3%) glucose (1.0%) medium (TYEG) + BCP indicator (0.04%).

2. 9 Durham tubes of sucrose (0.5%) + TYE broth + BCP (0.04%).

3. 9 Durham tubes of lactose (0.5%) + TYE broth + BCP (0.04%).

4. 9 Durham tubes of glucose (0.5%) + TYE broth + BCP (0.04%).

5. 9 Durham tubes of xylose (0.5%) + TYE broth + BCP (0.04%).

6. 1 tube each of pure cultures of *Enterobacter (Aerobacter) aerogenes* (1), *Pseudomonas aeruginosa* (2), *Alcaligenes viscolactis* (3), *Saccharomyces cerevisiae* (4), *Torulopsis sphericus* (5), your unknown culture (6), your partner's unknown (7), and *Lactobacillus arabinosus* (8). Note that cultures 4 and 5 are yeasts.

PROCEDURE

1. Label one Smith fermentation tube (see Fig. 15-1) as L.a. and the other as E.a. Hold the Smith fermentation tube in one hand with the tube portion pointing upward to avoid introduction of air bubbles. It is essential that no air be introduced into the tube portion since this would not only promote aerobiosis but would also obscure the release of gases produced by fermentation. Remember, only heterofermenters produce gases. If an air bubble enters into the tube, call your instructor for help in removing it before proceeding. Aseptically inoculate the tube labeled L.a. with 1.0 ml of *L. arabinosus* and similarly inoculate the tube labeled E.a. with *E. aerogenes*.

2. Incubate the Smith fermentation tubes in the 37C incubator for 2 days. Note: Smith fermentation tubes are easily tipped; exercise caution.

3. Obtain 9 Durham tubes of sucrose broth and label all of them with the letter S. Arrange the tubes in a row and label them 1 through 9. (Tube 9 will be the uninoculated control.) Aseptically inoculate tube 1 with 0.1 ml of culture 1 *(E. aerogenes)*. Mix the inoculum with the broth by rotating the tube between the palms of your hands. This is to insure that the inoculum will enter beneath the inserted Durham tube. Again avoid introduction of air bubbles. Similarly inoculate the remaining tubes with the cultures provided, including your unknown.

4. Repeat the same process with tubes of lactose, glucose, and xylose media.

5. Incubate all of the tubes in a 30C incubator for 2 to 5 days.

PERIOD II

MATERIALS

1. Frost gasometer.
2. A bottle of 40% KOH solution.
3. A bottle of methyl red indicator.
4. A piece of parafilm.
5. 1 glass stirring rod.

PROCEDURE

1. After incubation for 2 to 5 days, examine the tubes for growth of the cultures as indicated by turbidity of the medium, cell sediment at the bottom of the tube, formation of a ring or pellicle at the air and liquid interface. Also

look for a change in the color of the indicator, brom-cresol-purple (compare with uninoculated control tube). A yellow color indicates acid production (use symbol A for recording); pale yellow indicates production of slight acid (record this as SA); if there is no change in color (i.e., if the color is the same as that of the uninoculated control), record this as NC. If the indicator turned colorless, record this as R, meaning no acid was produced and the organism used the indicator as a hydrogen or electron acceptor. Remove the stopper from the tubes labeled 4 and 5 (yeast cultures) which show evidence of gas formation. Examine for odor. Is the odor familiar? Record your observations.

2. After examining the indicator change (change in pH of the medium), observe whether gas has been produced and collected in the Durham insert. Record your observations in the space provided. Use + for gas production and − for no gas formed.

3. After the gas production is recorded, measure the relative amount of gas produced by using the Frost gasometer as shown by your instructor (see Fig. 15-2). Record your measurements in the space provided.

4. *Qualitative analysis of the gas:* Carefully pour 40 per cent aqueous KOH solution into the tubes that contain gas until the tubes are 2/3 filled (the surface of the liquid should be 5 cm below the rim of the tube). If the KOH solution comes in contact with your hands or clothing, wash it away immediately with cold water as the reagent is extremely caustic.

5. Carefully mix the KOH solution into the medium with a glass stirring rod. The object of mixing is to promote entrance of the reagent into the

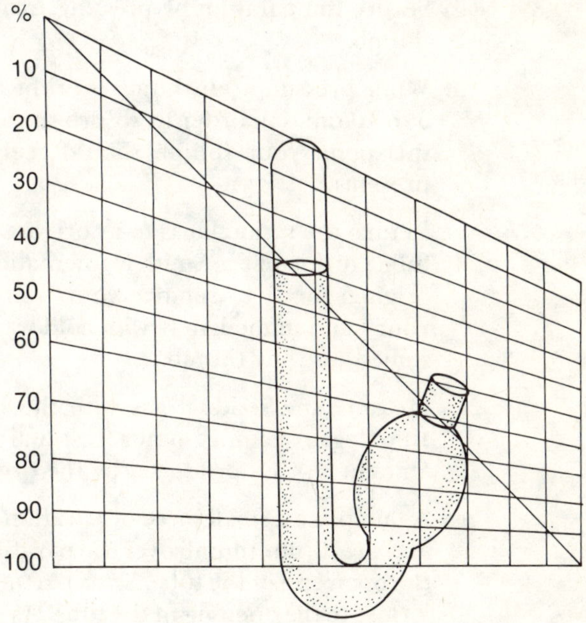

FIGURE 15–2. Smith Fermentation tube with gas aligned with Frost's gasometer. Example shown has 30 per cent gas.

Durham insert. If the gas accumulated at the top of the insert is CO_2, you will observe rapid disappearance of the gas as you stir the liquid.

$$CO_2 + 2KOH \longrightarrow K_2CO_3 + H_2O$$

If after mixing for 2 to 3 minutes, the gas bubble still remains in the tube, the residual gas contains little or no CO_2. For this experiment, consider the residual gas to be hydrogen. (Rinse stirring rod with cold water before you place it on the bench surface.) Measure the relative residual gas volume by using the Frost gasometer. Total gas volume minus residual gas volume gives the amount of CO_2 produced. Your instructor will explain which organism produced only CO_2, which organism produced both CO_2 and H_2, and the significance of the ratio of CO_2 to H_2. Record your results in the space provided.

6. Examine the Smith fermentation tubes for indicator color change. Measure the amount of gas produced with the Frost gasometer. Record your observations in the proper space provided. Remember, homofermenters will not produce gas.

7. Remove the cotton plugs and add 3 to 5 drops of methyl red indicator to each tube. Do not mix the indicator. Observe the change in the color of the methyl red indicator. If the indicator remains red, the pH of the medium is 4.5 or less; if it changes to yellow or orange, the pH of the medium is above 5. Your instructor will provide you with the readings from a pH meter for comparison. Record your observations.

8. a. Add 40 per cent KOH to the Smith fermentation tube with gas (the tube inoculated with *E. aerogenes*), almost to the rim of the tube.

   b. Place a piece of parafilm over the opening.

   c. Secure the parafilm by pressing firmly against the rim with your thumb.

   d. While pressing your thumb over the tube opening, invert the tube 5 to 10 times in order to absorb the CO_2 which may be present (at this point your thumb should feel a vacuum effect if CO_2 is present).

   e. Return the residual gas to its original position by gently tilting the tube portion of the Smith fermentation tube. After the gas is in the original position, remove your thumb from the opening. Wash your hand immediately with cold water whether or not KOH was spilled on your thumb.

   f. Measure the residual gas with the Frost gasometer and record. Initial gas volume minus residual gas volume represents the amount of CO_2 produced by the organism (see Fig. 15-2).

   g. Again place a small piece of parafilm over the opening of the tube and press your thumb over the parafilm. Return the residual gas to the open end of the tube. Your partner will light a match and bring it close to the opening of the tube. As you remove the film and your thumb, the flame should be brought over the tube opening. If the gas is hydrogen, it will explode with a small but distinctly audible "pop." Record your test results in the space provided.

NOTE: At the end of this experiment you have obtained the following information about your unknown in addition to that listed on pages 97 and 123:

1. Ability of the organism to ferment sucrose, lactose, glucose, and xylose.
2. Ability of the organism to produce gases from these sugars.

EXPERIMENT 16

Culture Techniques for Anaerobic Bacteria

In Experiment 11 we reviewed various cultural characteristics of bacteria and pointed out the differences between anaerobes and other types. Special techniques and media are employed for the growth of anaerobic bacteria and students should be familiar with them. Strict *(obligate)* anaerobes cannot grow in the presence of gaseous oxygen and even a brief exposure of vegetative cells to air may prove lethal. Two approaches have been used to grow obligate anaerobic bacteria: either oxygen must be excluded from the medium (environment), or the culture must be protected against the toxic effects of oxygen when it is not practical to rigorously exclude it.

There are several methods for excluding oxygen from the media. Bottles completely filled with broth which has been previously boiled to drive off dissolved oxygen may be used. Anaerobic bacteria can also be grown on conventional agar plates if they are incubated in special vacuum jars that can be evacuated and filled with pure hydrogen or nitrogen, or either gas mixed with CO_2. The Brewer anaerobic jar has a special lid with an electrically heated catalyst for removing the last traces of oxygen in the presence of hydrogen (Fig. 16-1). A more recent development by Brewer and Allgeier is the "gas pack." A disposable H_2 plus CO_2 generator envelope is placed inside an airtight jar containing a replaceable catalyst (Fig. 16-2). Water is added to activate the generator envelope and the jar is sealed. Oxygen in the system reacts with hydrogen to produce water in the presence of the catalyst, which is active at room temperature. A disposable anaerobic indicator (methylene blue) is often used along with the envelope to make certain that anaerobic conditions have been created and maintained. A completely disposable plastic bag, Gaspak (BioQuest), is also available for field work. The "gas pack" system eliminates the need for electrical power, vacuum pumps, tanks of gas, and pressure regulators. Another simple procedure, which works well for the cultivation of an occasional anaerobic culture, involves the use of pyrogallic acid. A plug of sterile cotton is shoved down an inoculated agar slant until the cotton almost touches the agar. Pyrogallic acid and sodium carbonate are then added, and the tube is tightly stoppered and inverted. All oxygen will be absorbed and CO_2 will be released.

The second approach to culturing anaerobic bacteria is to exclude as much air as possible and to rely on strong reducing agents to protect against the toxicity of the residual dissolved oxygen. Reducing agents such as sodium thioglycollate, cysteine, sodium formaldehyde sulfoxylate, ascorbate, or chunks of meat or liver have been used to lower the oxidation-reduction potential (Eh)

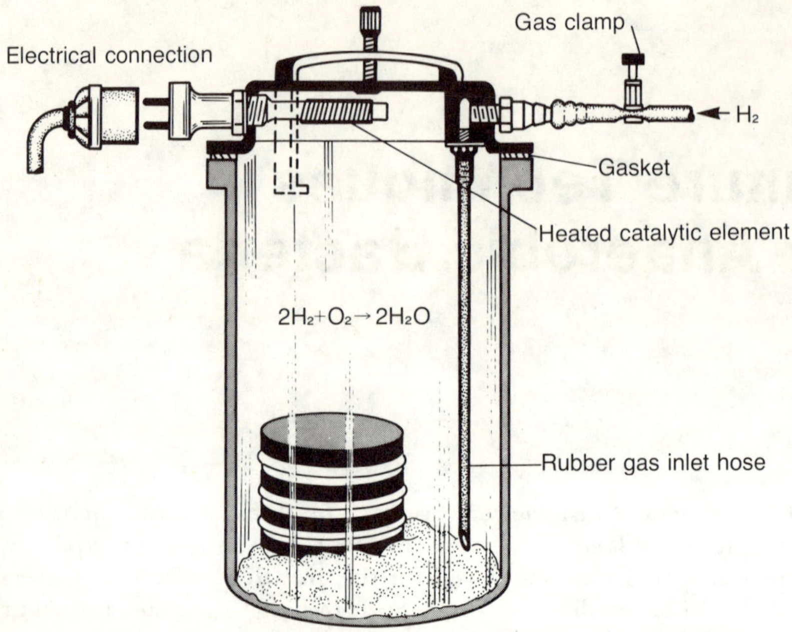

FIGURE 16–1. Cross section of a Brewer anaerobic jar showing the principal components.

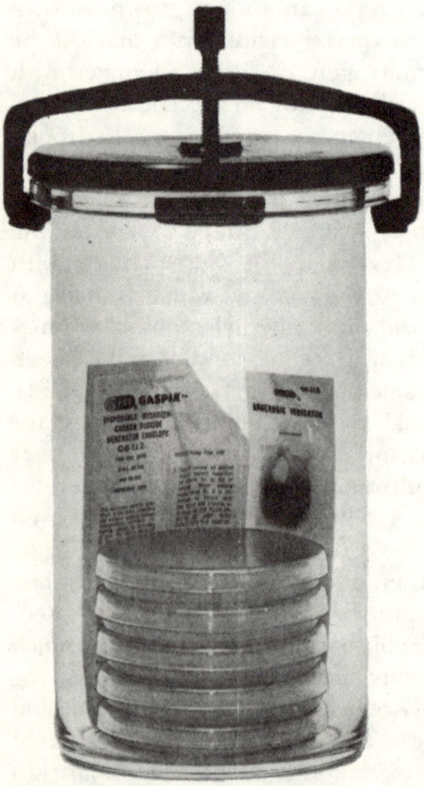

FIGURE 16–2. Simplified form of anaerobe jar utilizing hydrogen and carbon dioxide from the "Gaspak" envelope. Combination of hydrogen and oxygen in the jar is catalyzed at room temperature. (Courtesy of BioQuest, Division of Becton-Dickinson Laboratories, Baltimore, Md.)

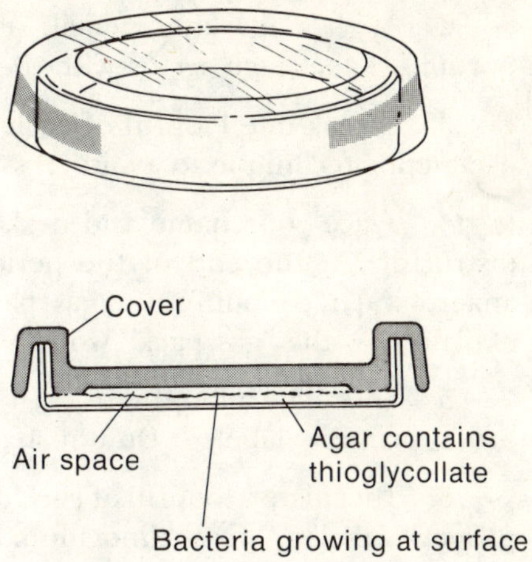

FIGURE 16–3. The Brewer anaerobic Petri dish cover in use. The agar contains the reducing agent, sodium thioglycollate. At the periphery of the plate, the cover is in contact with the agar, thus sealing the air space and excluding air.

of the media. Indicator dyes such as methylene blue or resazurin are often incorporated into the media since they are colorless in the reduced form and colored in the oxidized form. If the medium becomes colored, it means the Eh has risen well above the −100 to −300 mv considered desirable for the growth of most obligate anaerobes. Agar (0.05 to 0.2 per cent) is commonly added to such broths to increase viscosity, thereby reducing convection currents and mixing with the upper levels exposed to air. This helps to keep the deeper levels of broth anaerobic. A layer of sterile mineral oil at the surface also helps to exclude air. The "gas pack" has so simplified the handling of plated media that other methods have been largely abandoned; some laboratories, however, may still employ Brewer anaerobic covers for plates of thioglycollate agar (Fig. 16-3).

PERIOD I

MATERIALS

1. Nutrient broth cultures of *Bacillus subtilis* and *Staphylococcus aureus.*

2. Thioglycollate broth culture of *Clostridium sporogenes.*

3. 1 nutrient agar deep.

4. 3 tubes of nutrient broth.

5. 3 tubes of thioglycollate broth with indicator, Difco 0256 or equivalent (freshly boiled).

6. 1 sterile Petri dish.

7. Demonstration materials, if available, showing the variety of equipment used to cultivate anaerobic bacteria.

PROCEDURE

1. Pour the melted, cooled (45 to 50C) nutrient agar into a sterile Petri dish and allow to solidify.

2. With a marking pencil, sector the bottom of the plate into 3 equal portions. Label each sector with the name of one of the cultures provided.

3. Streak one loopful of each culture in the appropriately labeled sector; use aseptic technique to avoid cross contamination.

4. Place your name and desk number on the plate and hand in to your instructor. At the end of the period he will incubate all of the plates in an anaerobic jar containing a "gas pack." At that time he will demonstrate and explain how the "gas pack" works.

5. Inoculate a loopful of each culture into a separate tube of thioglycollate broth, properly labeled. Do not shake these tubes or create air bubbles.

6. Inoculate a loopful of each culture into a separate tube of nutrient broth, properly labeled. Mix the inoculum into the medium by rolling the tube between the palms of your hands.

7. Incubate all tubes at 30C for 2 to 4 days.

PERIOD II

PROCEDURE

1. Pick up your plate from your instructor, examine, and record your results.

2. Without shaking, examine the tubes of nutrient broth and thioglycollate broth. If growth (as evidenced by turbidity) is present, note whether it is at the bottom or top or whether it is throughout the tube. Record in space provided.

3. Time and archives permitting, your instructor will have on display various pieces (or photographs) of equipment used in the cultivation of anaerobes. Record for each the advantages and disadvantages.

RESULTS

1. Record the amount of growth as 0, +, + +, or + + +. Also indicate whether growth was at top, bottom, or throughout.

	Species of Bacteria		
Medium	*C. sporogenes*	*B. subtilis*	*S. aureus*
Nutrient broth (aerobic)			
Thioglycollate broth (anaerobic)			
Nutrient agar plate incubated anaerobically			

DISCUSSION

1. What did the color change at the top of the thioglycollate medium indicate?

2. On the basis of your results, how would you classify each of the three bacterial species, aerobic, anaerobic, or facultative?

all three

3. Where might you expect to find anaerobic bacteria in nature?

4. Can you name disease-causing anaerobic bacteria of particular concern to the food industry?

botul

EXPERIMENT 17

Measurement of Bacterial Growth

In reference to multicellular organisms, growth usually means an increase in the size of the individual. However, when we speak of the growth of unicellular organisms, such as bacteria, we generally mean an increase in the number of individuals, or its corollary, an increase in the cell mass. Bacterial growth may be measured in a variety of ways, some of which have certain advantages for particular applications. Increased mass of a culture can be determined by sedimenting the cells and noting the increase in packed cell volume or the increase in either wet or dry cell weight. Or one may measure the increase in a particular cell component (such as nitrogen, protein, or nucleic acid), or measure a product of metabolism such as an acid. The easiest and most widely used procedure, however, is to measure the increase in turbidity of the broth culture over a period of time using a spectrophotometer, colorimeter, or nephelometer (Fig. 17-1). The greater the number of cells, the greater will be the turbidity (light scattering). One of the major disadvantages of this method is that it works well only in the range of 10^6 to 10^{10} bacteria per ml for most species.

For many applications it is desirable to determine the number of cells present. Direct microscopic counts can be made using special counting chambers (Petroff-Hauser chamber) or stained slide preparations with a known volume spread over a known area (e.g., special slides used for direct counts on milk). These techniques do not allow one to differentiate between living and dead cells. (Can you think of a situation in which it might be helpful to know the total bacterial count, both living and dead?) The number of viable cells can be estimated through simple serial dilution (Experiment 9; also see Most Probable Number, Experiment 29), but is most frequently determined by plate counts. This assumes that when sufficiently low numbers of bacteria are placed on the surface of an appropriate medium, each viable bacterium will replicate and give rise to a single, well separated colony. This is not necessarily true, but the results are generally so close to the ideal that the technique is acceptable for most purposes. Organisms such as staphylococci, lactobacilli, and capsule-producing bacteria do present a special problem because these organisms have a tendency to aggregate, or form clumps or chains. Here the total microscopic counts are invariably higher than the colony count and one frequently sees the term "colony-forming units" as an acknowledgement of this fact. A brief exposure to sonic oscillation may break the clumps or chains apart and will allow a much more accurate count to be made.

INSTRUMENTS USED TO MEASURE TURBIDITY (OPTICAL DENSITY)

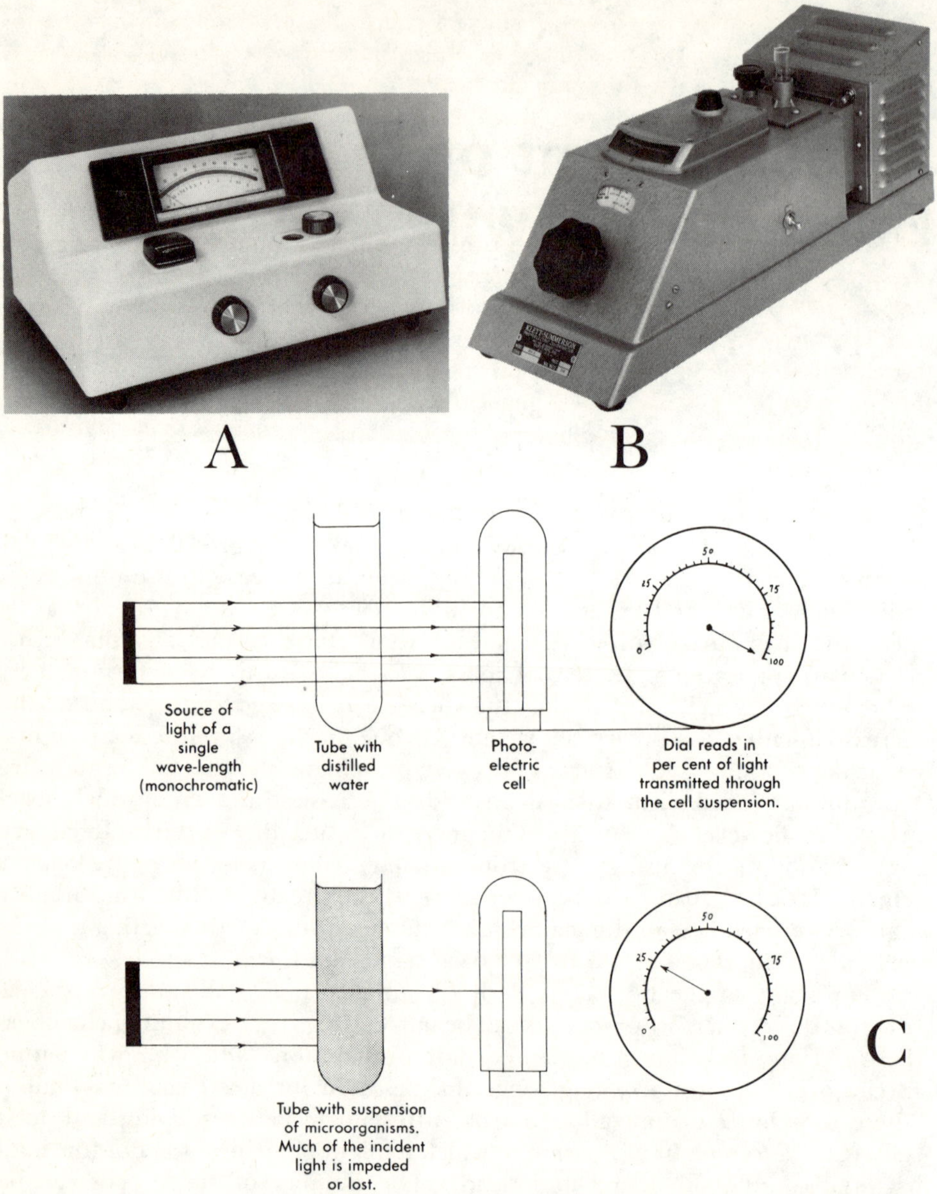

FIGURE 17–1. *A,* Spectrophotometer (courtesy of Bausch and Lomb). *B,* Colorimeter (courtesy of Klett-Summerson). *C,* Schematic of photocolorimeter (from Frobisher, M., et al. *Fundamentals of Microbiology,* 9th ed. Philadelphia: W. B. Saunders Co., 1974).

A typical bacterial cell in a favorable environment grows by increasing in size about two-fold, then dividing to form two daughter cells *(binary fission).* The two daughter cells divide to form four cells; these divide to form eight, and so on. The time between cell divisions, called the *generation time,* may be many hours or less than 15 minutes, depending on the species and cultural conditions.

When the cells are inoculated into fresh medium, they will not divide immediately but actively synthesize materials necessary for their growth from the new environment. This period, while the organisms are physiologically and biochemically active but show no increase in number, is referred to as the *lag phase.* Once division has begun, the bacteria will continue to divide with the shortest generation time possible in their particular environment. So long as nutrients are plentiful, the numbers will increase exponentially *(logarithmic growth phase).* Eventually, some required nutrient will be depleted, gaseous exchange will be insufficient, or a toxic waste product will begin to accumulate, and the viable cell population will remain the same. This period is called the *maximum stationary phase.* If conditions remain unfavorable for growth for very long, large numbers of cells may begin to die at an exponential rate *(death phase).* To see that this is indeed a fortunate thing for our planet, calculate the number of cells and the cell mass of a culture with a generation time of 20 minutes resulting from continuous logarithmic growth over a 48-hour period. (Assume each cell weighs 1×10^{-12} grams; for comparison, the earth's mass is roughly 5×10^{30} grams.)

In this experiment you will trace the growth of a culture over a period of time; that is, you will construct a growth curve by determining viable counts. In addition, turbidity of the culture will be followed and the two sets of data will be compared. Time and facilities will not permit each student to follow growth over a 24-hour period; therefore, this will be a class experiment whereby each student (or pair) will obtain the required data for a particular time. The class results will then be pooled so that each student can construct a growth curve and calculate the generation time. Recall that:

$$G = \frac{t}{n} = \frac{t}{3.3 \log b/B}$$

where
G = generation time
t = time period ($t_2 - t_1$)
n = number of generations
B = number of bacteria at beginning of time period (t_1)
b = number of bacteria at end of time period (t_2)

Also remember that generation times must be based on the logarithmic or exponential growth portion of the curve. Why?

PERIOD I

MATERIALS

The following will be required for each sample:

1. 1 bottle of nutrient agar (150 ml).
2. 6 sterile Petri dishes.
3. 2 bottles of sterile water (99 ml).
4. 2 tubes of sterile water (9 ml).
5. 150 ml beaker for 95% alcohol.

6. 1 bent glass spreading rod.

7. Canister of pipettes (1 ml and 0.1 ml).

PROCEDURE

1. The instructor (or teaching assistant) will inoculate a large shaker flask containing 500 ml of nutrient broth with 1.0 ml of a freshly grown broth culture of *Escherichia coli* (1×10^8 cells/ml) some 10 hours before class. The culture will be grown at 37C with aeration; 5 ml samples will be removed aseptically at the intervals shown below, placed in sterile tubes, and immediately chilled in an ice bath to stop growth. These tubes will be kept chilled until students pick them up at the beginning of the regularly scheduled laboratory period.

NOTE: Where facilities exist, students may wish to grow the culture themselves. A sign-up sheet will indicate which student(s) must take a sample at the indicated time. Under these conditions, turbidity readings and plate counts will be made immediately after taking the sample. While this procedure is preferable, scheduling and security may prevent its use in some laboratories.

2. 30 samples will be collected over a 24-hour period, or your instructor may wish to concentrate on only the first 10 hours of growth (15 samples) and assist you by providing representative data, based on experiments in previous years, for the remaining 14 hours. See the Results section for recommended sampling times and dilutions.

3. Pour the melted, cooled (45 to 50C) nutrient agar into 6 sterile Petri dishes, distributing it equally. Avoid air bubbles. Allow agar to solidify fully before using plates.

4. Take the culture sample provided and determine the optical density. Follow the directions adjacent to the instrument or given by the instructor. Be sure you understand how the instrument works before attempting to measure your sample. Record your results in the appropriate space.

5. Some workers prefer to record transmittance, then convert values to optical density (O.D.) by use of a table or formula (O.D. = 2−log of % transmittance). Why might this result in greater accuracy?

6. From the schedule (Results section), determine what 3 *final* dilutions must be made for your particular sample. Then follow the scheme below, plating in duplicate. Be sure to label your plates with time period and *final* dilution. Accurate pipetting is essential!

7. Aseptically add 0.1 ml of the sample to the surface of each plate.

   a. Place the plate on the turntable.

   b. Dip the bent glass spreading rod into 95 per cent alcohol provided in a beaker. Burn off excess alcohol by quickly passing the rod over the Bunsen burner flame. This is to sterilize the rod. (Do not leave the rod in the flame; glass rods will break easily.) Repeat alcohol sterilization (see Fig. 17-2).

PROPER SPREADING TECHNIQUE

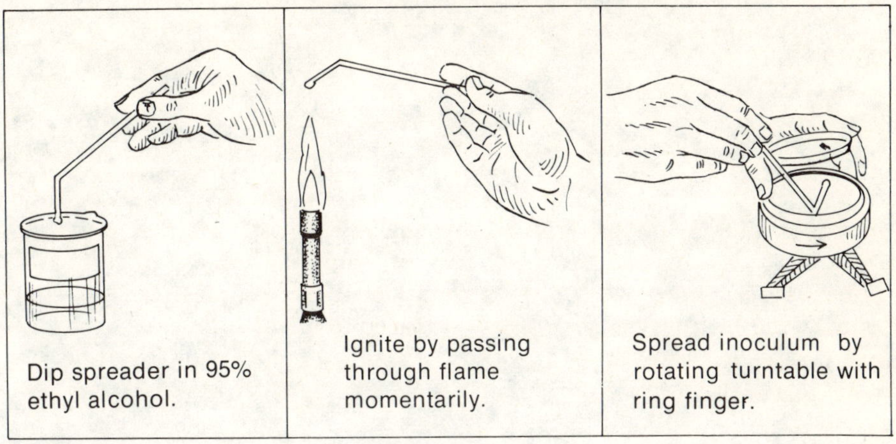

FIGURE 17–2

c. Spread the culture with the sterilized glass rod as shown by your instructor. Be sure to sterilize the rod between plates.

8. Invert the plates and incubate at 30C for 2 days.

PERIOD II

MATERIALS

1. Semi-logarithmic graph paper.

PROCEDURE

1. If a colony counter is available, use it as an aid and count the number of colonies on each plate. Remember, the most accurate range of counts is 30 to 300. Many students find it easier to examine the plates bottom up and mark each colony with the marking pencil as it is counted.

2. Select the set of duplicate plates with counts in the appropriate range, average the counts, and calculate the number of bacteria per ml of sample. Enter your results.

3. After the class data have been pooled, plot optical density and logarithm of viable bacteria vs time of growth on the semi-logarithmic graph paper provided. See sample graphs.

4. Identify the major phases of growth: lag, exponential, stationary, and death.

5. Calculate the generation time for the culture.

RESULTS

1. Record your results and those of your classmates in the proper space.

Sample (hrs)	Optical Density	Recommended Final Dilutions (log)	Average of Best Counts	Bacteria/ml Sample
0		−3, −4, −5		
1		" " "		
2		" " "		
2.5		−4, −5, −6		
3		" " "		
3.5		" " "		
4		" " "		
4.5		−5, −6, −7		
5		" " "		
5.5		" " "		
6		−6, −7, −8		
6.5		" " "		
7		" " "		
8		" " "		
10		" " "		
12		" " "		
14		" " "		
16		−5, −6, −7		
18		" " "		
20		" " "		
22		" " "		
24		−4, −5, −6		

2. The generation time of *E. coli* grown under these conditions is ________.

DISCUSSION

1. If *E. coli* were grown again under the same conditions, could you predict the number of viable cells per ml merely by measuring the optical density of the culture? In all phases of growth or only in some? What if another type of medium were selected?

2. For some types of work it is necessary to keep a bacterial culture in the exponential growth phase indefinitely. What type of apparatus is used for this purpose?

3. What causes the stationary phase?

SECTION

III

INHIBITION AND KILLING OF MICROORGANISMS

EXPERIMENT 18

Activities of Selected Antibiotics

An antibiotic is defined as a metabolic product of one organism that in very small amounts is detrimental or inhibitory to other microorganism(s). Antibiotic substances are produced by a wide range of living organisms, from man to plants to microbes. The production of antibiotics by microorganisms is readily demonstrated in the laboratory. Bacteria of the genera *Bacillus* (polymyxin B) and *Streptomyces* (streptomycin), and a common genus of mold, *Penicillium* (penicillin), all produce antibiotic substances, as indicated, that diffuse into the culture media in which they are growing.

Although there are many antibiotics commercially available today, penicillins are still considered the most effective and nontoxic for treatment of various bacterial diseases of man and other animals. An antibiotic which has the ability to inhibit or destroy many different kinds of pathogenic microorganisms is called a *broad spectrum* antibiotic. The antimicrobial mechanism of some antibiotics is their ability to serve as structural analogues of certain essential cell components or enzymes. For example, penicillin inhibits transpeptidation in the final step of cell wall formation, that is, the formation of cross links between polypeptide side chains. Presumably, this happens because a portion of the penicillin molecule is a structural analogue of D-alanyl-D-alanine and combines irreversibly with the enzyme required for cross linking (see text for further discussion). Streptomycin (a broad spectrum antibiotic particularly active against gram-negative organisms, but now less frequently used) appears to inhibit the enzyme(s) involved in the early stages of cell protein synthesis. On the other hand, some antibiotics, such as bacitracin, may act as surfactants and attack the cell membrane, possibly by destruction of lipid constituents of the membrane and cell wall.

The purpose of this experiment is to demonstrate the production of penicillin and streptomycin and the activities of these as well as commercially prepared antibiotics.

PERIOD I

MATERIALS

1. Pure cultures of *Penicillium chrysogenum* (or *P. notatum*) and *Streptomyces griseus* on agar slants.

2. 2 sterile plates containing antibiotic assay agar (tryptone, 0.5%; peptone, 0.5%; yeast extract, 0.3%; beef extract, 0.1%; glucose, 0.2%; agar, 1.5%; and 1000 ml distilled water).

3. 1 tube of nutrient broth.

PROCEDURE

1. Aseptically transfer one loopful of *P. chrysogenum* (or *P. notatum*) to the center of one of the assay agar plates. Then gently, without cutting into the agar, spread the inoculum in a single, narrow line so as to divide the plate into equal halves with the inoculum (see Fig. 18-1).

2. Do the same on the other plate with the *S. griseus* culture.

3. Invert the plates and incubate in your desk for 4 to 7 days.

4. Inoculate your unknown into the nutrient broth provided and incubate it in your desk until next period.

PERIOD II

MATERIALS

1. Pure cultures (24 hr) of *Bacillus cereus* (1), *Escherichia coli* (2), *Staphylococcus aureus* (3), *Pseudomonas aeruginosa* (4), and your unknown culture (5) in tubes of nutrient broth.

2. 3 sterile nutrient agar plates.

3. 3 tubes containing 3.0 ml of melted, sterile antibiotic assay soft agar.

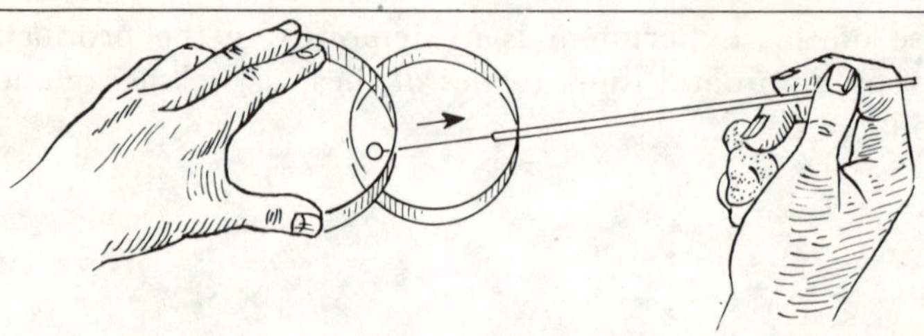

Prepare streak culture of either *P. chrysogenum* (or *P. notatum*) or *S. griseus* on an antibiotic assay agar plate.

FIGURE 18–1. Streak plate assay of penicillin and streptomycin.

(The constituents of assay soft agar are identical to those of the assay agar used in Period I except that the agar content is reduced to 0.7%.)

4. Various commercially prepared antibiotic disks.

PROCEDURE

1. Select 2 cultures, one gram-positive and the other gram-negative, from the 4 cultures provided in addition to your unknown. Your partner will use the other 2 cultures and his unknown. (Cultures 1 and 3 are gram-positive and 2 and 4 are gram-negative.)

2. Obtain 3 tubes of the antibiotic assay soft agar from the water bath (kept at 45 to 50C).

3. Aseptically pipette 0.2 ml of your first culture into one of the tubes of melted soft agar. Mix thoroughly by rotating the tube between your palms. Pour the entire contents into one of the nutrient agar plates. Spread the inoculated agar evenly by tilting the plate back and forth. Allow to solidify. Do the same for your second culture and your unknown. Label the plate with the name of the culture used. Your partner will do the same with the other two cultures and his unknown.

4. Take out the *P. chrysogenum* (or *P. notatum*) and *S. griseus* plates from your desk. These cultures should be sporulated by this time. Examine and record your observations as to amount of growth. Invert the plates and with your marking pencil draw 5 lines at least 3 cm long and 1 cm apart perpendicular to the line of growth. (See Fig. 18-2.) Avoid the extreme ends of the growth line. Label the lines with 1, 2, 3, 4, and 5 (unknown). These lines will be used as a guide in the next step.

5. Aseptically transfer a loopful of culture 1 onto the midpart of line 1. Spread the culture toward the edge of the plate and toward the growth of

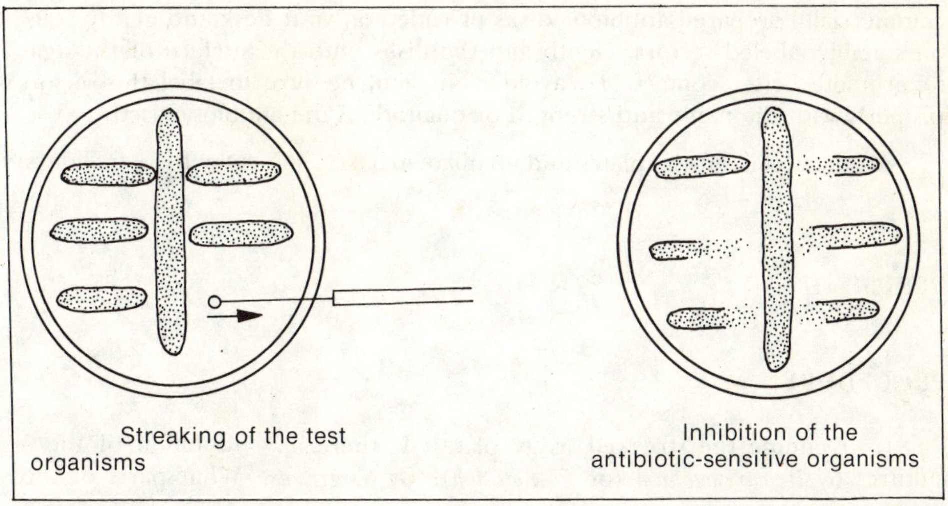

FIGURE 18–2. Antibiotic sensitivity test of known and unknown cultures by streak plate assay.

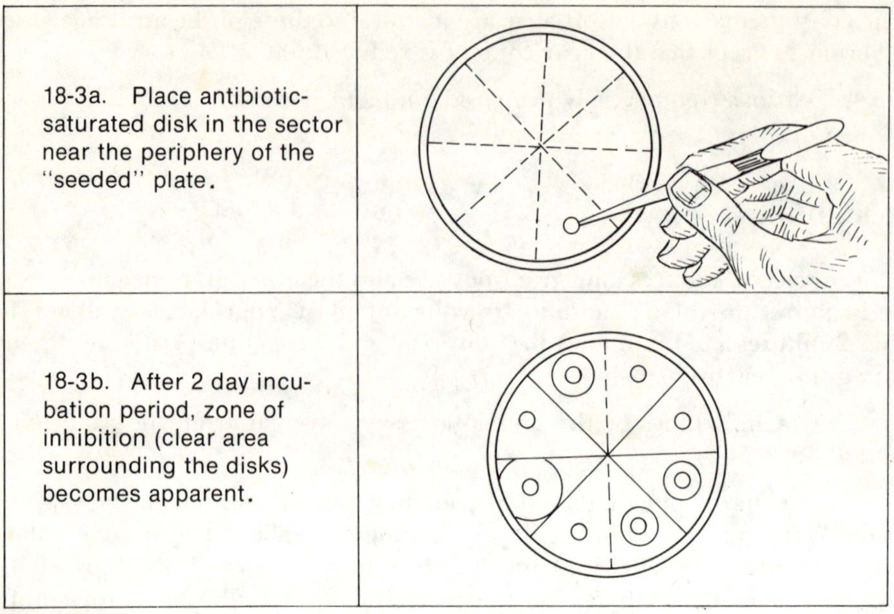

FIGURE 18–3. Antibiotic assay by disk or tablet method.

P. chrysogenum (or *P. notatum*) and *S. griseus*. Be careful not to touch the growth. Flame the loop and do the same for the other cultures—culture 2 on line 2, etc.—until all 5 cultures are streaked on both plates (Fig. 18-2a). (If these two organisms produced antibiotics which diffused into the agar, we anticipate that an antibiotic-sensitive culture streaked near the growth should be inhibited. See Figure 18-2b.)

6. After the soft agar (seeded in Step 3) solidifies, invert the plates and sector the bottom of each into 8 equal parts with a marking pencil. Number the sectors 1 through 8. Set the plates side by side for the next step (Fig. 18-3).

7. With sterilized forceps (dip into alcohol and ignite), aseptically pick up commercially prepared antibiotic disks provided on your desk and gently place on suitably labeled sectors. Gently tap the disks onto the surface of the agar to promote better contact. To avoid confusion, be sure to label the sectors properly with the name and strength or quantity of the antibiotic used.

8. Gently invert the plates and incubate in a 37C incubator for 1 to 2 days.

PERIOD III

PROCEDURE

1. Examine the streaked assay plates. Is there any inhibition of the 5 cultures by *P. chrysogenum* (or *P. notatum*)? By *S. griseus*? What parts of the

streaked cultures are inhibited? Observe carefully and record your observations in the space provided.

2. Observe the disk assay plates. Measure the radii of the clear zones formed by the antibiotics in mm and record your results in the space provided.

3. Exchange disk assay plates with your partner and record your observations in the space provided.

2 N.A. deeps
2 sterile plates
1 Known M.O.
1 Unknown

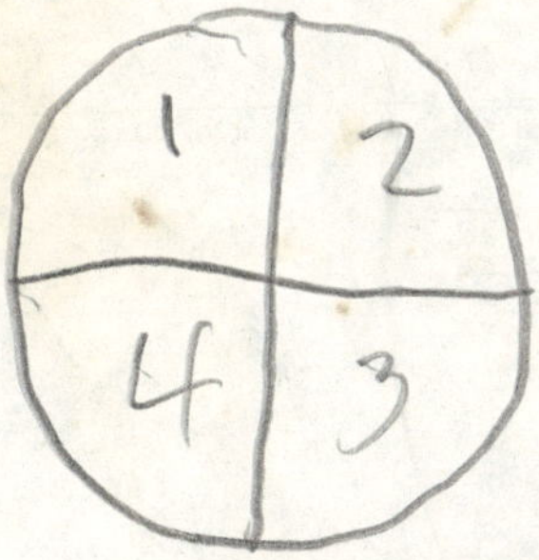

1 Kanamycin
2 Neomycin
3 Novobiocin
4 Erythrizomycin

EXPERIMENT 19

Effects of Physical Agents

Controlling microbial populations often involves the use of various physical agents. Some of the more commonly used agents are low temperature, incineration, high temperature (either dry or moist heat), osmotic pressure, ultraviolet light, x- and gamma rays, and sonic and ultrasonic waves. These agents are used primarily to destroy microorganisms *in* or *on things,* including bacteriological media. Under proper conditions, their effects on the population are considered irreversible: microorganisms cease to reproduce even under a return to their ideal environment. The action of the agents may depend on destruction or denaturation of essential cell components, such as enzymes (proteins) and nucleic acids, or on mechanical rupture of the cell walls and membranes to a point at which all reactivation and restoration are impossible. In this experiment, the effects of temperature and ultraviolet light on the survival of microorganisms will be studied.

Temperatures above a "maximum" will exert a killing effect, whereas temperatures below a "minimum" are regarded as having a "static" effect. Lethal action of heat on microorganisms is affected by numerous factors: (1) intensity and kind of heat, (2) duration of contact, (3) history of microorganisms (i.e., age of vegetative cells or endospores), (4) number of organisms, and (5) pH of the medium in which the organisms are present. Moist heat, such as steam sterilization in an autoclave, presumably denatures the cytoplasmic membrane and subsequently destroys other proteins and nucleic acids. Bacterial endospores, the most resistant microorganisms, are killed at a much lower temperature and in a shorter time by moist heat than by dry heat. Dry heat can destroy microorganisms through dehydration and oxidation or even incineration. A good example of the latter is your practice of flaming wire inoculating needles and loops to render them sterile.

The susceptibility of different organisms to high temperature can be measured by two methods: thermal death point (TDP, now less frequently used), which is the lowest temperature at which a suspension of bacteria is killed in 10 minutes; or thermal death time (TDT, more specific than TDP), which is the time required to kill a given number of cells or spores at a given temperature. The temperature selected for pasteurization of milk is based on the TDT of *Coxiella burneti,* the cause of Q fever, which may be present in milk and is known to be more heat resistant than *Mycobacterium tuberculosis,* another pathogen which can be transmitted by milk. Thus pasteurization is not a sterilization process but a process which kills microorganisms of certain types, the pathogens; pasteurized milk may still spoil when left standing at room temperature for any period of time.

In this experiment we will subject to various temperatures raw milk containing both indigenous flora and *Serratia marcescens, Sarcina lutea,* and sporulated *Bacillus cereus* (or *B. megaterium*) as indicator organisms.

Ionizing radiation, such as gamma, x-, and cathode rays, and the longer electromagnetic wave radiations, such as ultraviolet rays, have lethal effects on microorganisms. The wave length of UV ranges from 15 to 340 nm, with the 265 nm region considered to have the highest bactericidal efficiency. For this reason, UV lamps having wave lengths between 260 and 270 nm are widely used to reduce the microbial populations in hospital operating rooms and aseptic filling rooms in the pharmaceutical industry, or even in culture transfer rooms when they are not in use. The mechanisms by which ultraviolet radiation kills microorganisms are not completely understood. Nucleic acids and to some extent proteins, the essential constituents of living matter, exhibit very strong absorption of ultraviolet radiation; hence the indications are that cellular damage involves change in nuclear substance. It has been hypothesized that the detrimental effect may be accomplished by oxidation of certain substances after the formation of peroxides.

The germicidal effect of ultraviolet radiation on microorganisms can be reduced by immediate exposure to an intense source of visible light. This apparent reversal of the lethal action of UV radiation by visible light is called *photoreactivation.* Depending upon the dosage, ultraviolet radiation may be used as a mutagenic agent.

PERIOD I

PART A. INHIBITING AND KILLING OF MICROORGANISMS BY APPLICATION OF HEAT

MATERIALS

1. 1 flask of raw milk containing *Serratia marcescens, Sarcina lutea,* and *Bacillus cereus* (or *B. megaterium*).
2. 4 sterile test tubes with metal caps.
3. 1 bottle of melted nutrient agar (approximately 80 ml).
4. 4 sterile Petri dishes.

PROCEDURE

1. Dispense the milk into the 4 sterile test tubes with metal caps provided (each 1/3 full). Try not to run the milk down the wall of the tubes.

2. Label the first tube as untreated, the second tube as pasteurized, the third tube as 100C for 3 min, and the fourth tube as sterilized. Put your initials on the tubes and submit the fourth tube for sterilizing in an autoclave.

3. Aseptically pour melted nutrient agar into the 4 sterile Petri dishes provided; allow to solidify.

4. Using a sterilized wire loop, prepare a streak plate from the untreated control using one of the nutrient agar plates prepared. Label the plate as untreated control.

5. a. Place the second tube, labeled as pasteurized, into a water bath maintained at 63C for 30 minutes. At the end of 30 minutes, remove the tube and cool under running tap water for 1 minute.

   b. Remove 1 loopful of the pasteurized milk with a sterile wire loop and streak it on the second plate. Label the plate as pasteurized.

6. a. Place the third tube into a beaker of gently boiling water (100C) for 3 minutes. Note that any organisms on the wall of the tube above the water line will receive less heat treatment. Cool the contents under running tap water.

   b. Prepare streak plate as in Step 5b.

7. a. Obtain your tube of sterilized milk from your instructor (from Step 2). The milk has been autoclaved under 15 pounds of steam pressure (121C) for 15 minutes and is considered sterile, that is, free of all living things, including viruses.

   b. Prepare a streak plate as in Step 5b.

8. Prepare a smear on a glass slide from the untreated control. Air dry and heat fix. Wash the smear with xylene (dropwise) to remove any fats which may be present in the heat-fixed smear. (The slide should be tilted over the bench sink to receive the xylene.) Five to 10 drops of xylene will serve the purpose. Allow to air dry and stain 2 minutes with aqueous methylene blue. Place the slide in your slide box and save until Period II.

9. Incubate all the material in your desk cabinet for 2 to 5 days.

PART B. EFFECTS OF ULTRAVIOLET RADIATION ON THE MICROORGANISMS

MATERIALS (For a pair of students.)

1. 1 bottle of melted nutrient agar (100 ml).
2. 5 sterile Petri dishes.
3. Broth cultures of *S. marcescens* and *B. cereus* (sporulating).
4. 2 5×8 file cards.
5. A canister of pipettes.

6. A roll of aluminum foil.
7. Ultraviolet lamp (260 nm).
8. Photoflood lamp.
9. Glass spreading rod, alcohol, turntable.

PROCEDURE

1. Aseptically pour melted nutrient agar into each of 5 sterile Petri dishes. Allow to solidify.

2. Draw any pattern (triangle, square, circle, a letter E, V, etc.) approximately 4 cm square in the center of a 5×8 card. Cut out the pattern with scissors. You will need two patterns, one for each culture.

3. Label the first plate as S.m. 15 sec, the second plate as B.c. 30 sec, the third plate as S.m. 30 sec, the fourth plate as B.c. 1 min, and the last plate as S.m. 15 sec + photoflood lamp 1 min.

4. Aseptically add 1 drop of *S. marcescens* onto each of the 3 plates labeled S.m. Spread the culture with a sterilized glass rod (refer to Experiment 17 and Fig. 17-1).

5. Repeat Step 4 using *B. cereus.*

CAUTION: Do not look directly at the ultraviolet light. UV light may cause permanent damage to your eyes.

6. a. Remove the lid of the plate labeled S.m. 15 sec and immediately place the card pattern over the plate. (Make certain that the pattern you cut out is approximately in the center of the seeded lawn.)
   b. Place the plate with the pattern directly under the UV lamp (see Fig. 19-1).
   c. Turn on the lamp for the prescribed time, remove the card and place on next plate (Step d), then replace the Petri plate lid. Place a piece of aluminum foil over the plate to avoid excessive exposure to daylight.

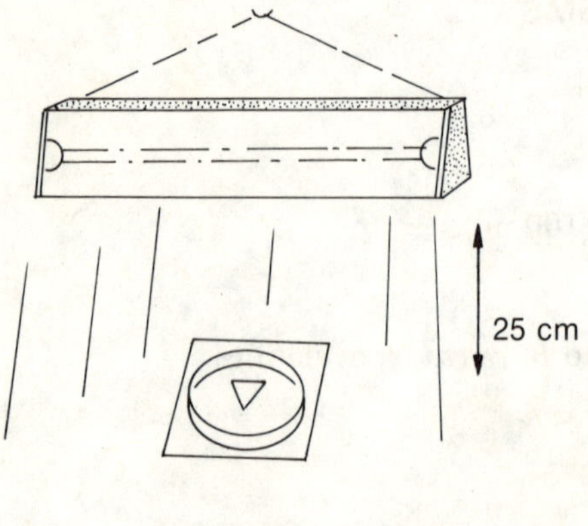

FIGURE 19–1. Exposing the culture to UV light.

d. Repeat the process with plates labeled S.m. 30 sec, B.c. 30 sec, B.c. 1 min. Avoid cross contamination of cultures by using the second card for *B. cereus.* Stack all plates exposed to UV and wrap them in foil. Invert and incubate at 30C until next period.

7. a. Remove the lid of the plate labeled S.m. 15 sec + photoflood lamp 1 min and immediately place the card and treat the plate as in Steps 6a to c; however, do not use the aluminum foil.

b. Remove the card and place the irradiated plate directly under the photoflood lamp for 1 minute. Replace the lid and incubate the plate in an inverted position without the aluminum foil cover until the next period.

PERIOD II

PROCEDURE

PART A. INHIBITING AND KILLING OF MICROORGANISMS BY APPLICATION OF HEAT

1. Observe the changes which took place in each of the 4 tubes of milk and describe the appearance of each tube in the space provided.

2. Observe the streak plates and record the kind of colonies which grew on the surface. Note particularly the color of the colonies. Record your observations in the space provided.

3. Make methylene blue stained smears from untreated, pasteurized, and boiled milk as in Step 8 of Period I. You may place all 3 smears on the same glass slide provided the smears do not touch each other.

4. Observe the smears prepared in Period I under the oil immersion objective and record your results in the space provided. Do the same for the slide prepared in Period II. Record your observations in the space provided.

PART B. EFFECTS OF ULTRAVIOLET RADIATION ON THE MICROORGANISMS

1. Examine the UV irradiated plates and compare the results in terms of:

   a. Difference in sensitivity between *S. marcescens,* nonsporeformer, and *B. cereus,* sporeformer.

   b. Lethal effect of UV irradiation with time of exposure for both organisms.
   Record your observations in the space provided.

2. Observe the photoreactivated plate and compare it with the plate labeled S.m. 15 sec. Record your observations.

DISCUSSION

1. Based on your experience (Part A), how would you isolate spore-forming microorganisms from soil?

2. What are some of the reasons that sporeformers withstood pasteurization and treatment such as boiling for 3 minutes?

3. What effect would you expect if the distance between the UV lamp and surface of the seeded plate is increased?

EXPERIMENT 20

Effects of Chemical Agents

There are extensive numbers of *disinfectants* (chemical substances that inhibit or kill microorganisms and are usually used on inanimate surfaces) and *chemotherapeutic agents* (chemical substances usually used in treatment of infections caused by microorganisms). See your text for a review of the various agents and descriptive terms. No single chemical or antimicrobial agent, however, is "best" or "ideal" for all purposes. Certain types of compounds are particularly effective in some situations and of little or no value in others. Users of disinfectants or chemotherapeutic agents must always consider (1) the kind of material to be treated, and (2) the type of microorganism to be controlled. In addition to these two factors, general conditions such as temperature, pH, contact time, and concentration of the agent must be considered. If the agent's lethal action is oxidation of cell constituents, then the presence or absence of extraneous organic material may have a bearing on the rate and efficiency of inhibition or killing of the microorganisms.

There are many methods of evaluating the disinfectant power of chemicals, but the *phenol coefficient method* is the oldest and the most frequently employed in evaluating other chemical compounds that may have inhibitory or killing action on microorganisms. The phenol coefficient is a figure comparing the dilutions of phenol and of another chemical that possesses equivalent killing power for a specified test organism, usually *Staphylococcus aureus* (ATCC 6538) or *Salmonella typhi.* For example, if a 1:90 dilution of phenol solution completely destroys a certain number of *S. aureus* organisms in 10 minutes, and chemical compound X diluted to 1:900 completely destroys the same number of organisms in 10 minutes, this means that chemical X can be diluted 10 times as much as phenol and still possess equivalent killing power for the culture under *in vitro* conditions. Details of the phenol coefficient method and calculation of phenol coefficient are described in Table 20-1. Phenol is bactericidal; it denatures and precipitates cell proteins.

Another commonly used and somewhat simpler method is the *agar plate test.* The culture (or cultures) is seeded on an appropriate agar medium and filter paper disks saturated with various dilutions of the chemical or different chemicals of the same dilution are placed on the seeded plate. After the plates are incubated, zones of inhibition of the bacterial growth surround those concentrations of substances that possess bactericidal or bacteriostatic action. The diameter of the clear zone (inhibited zone) depends upon the diffusibility of the chemicals and their antibacterial potency.

TABLE 20-1
TYPICAL DATA FROM A PHENOL COEFFICIENT DETERMINATION

Disinfectants	Dilutions	5 min Sub-cultures	10 min Sub-cultures	15 min Sub-cultures
Phenol	1:75	−*	−	−
	1:80	−	−	−
	1:90	+*	−	−
	1:100	+	+	+
"Unknown"	1:350	−	−	−
	1:400	+	−	−
	1:450	+	−	−
	1:500	+	+	−
	1:550	+	+	−
	1:600	+	+	+
	1:650	+	+	+

* − = absence of growth; + = growth

NOTE: Testing conditions must be identical for both phenol and the unknown chemical. You must employ standard test cultures and the test must be performed at 20C. After addition of the organism into the serially diluted phenol and unknown reagent, all the diluents are placed in a water bath set at 20C and subcultures are made from each tube after 5 min, 10 min, and 15 min. The phenol coefficient is then calculated as the ratio of the highest dilution of "unknown" not killing the organism in 5 min (evidenced by growth in the corresponding broth tube), but killing in 10 min (evidenced by absence of growth in the broth), to the corresponding dilutions of phenol. The values obtained are shown above. The 1:90 dilution of phenol failed to kill in 5 min but did so in 10 min; the 1:450 dilution of unknown did the same. The ratio of unknown to phenol is $\frac{450}{90}$, or 5

$$\text{Phenol coefficient} = \frac{\text{dilution of X that kills test organism in 10 min, but not 5 min}}{\text{dilution of phenol that kills test organism in 10 min, but not 5 min}}$$

Bacteriostatic agents (substances which prevent the growth of microorganisms) and *bactericidal agents* (substances which kill microorganisms) usually fall into two categories: (1) chemicals that injure cell membranes (synthetic detergents), and (2) chemicals that inhibit enzyme activity or denature proteins. Inhibition of essential coenzymes or enzymes may be reversible or permanent, in which case the cell will be killed. Heavy metals, such as $HgCl_2$ and $AgNO_3$, often react with enzymes or protein amino acid radicals such as $-NH_2$ and $-SH$, thereby causing inactivation of the essential enzymes and proteins. Some antibacterial agents may act as structural analogues (compounds with very similar molecular configuration) of essential growth substances. The sulfur drug sulfanilamide is a structural analogue of para-aminobenzoic acid, a precursor compound for the synthesis of folic acid, a growth factor required by many microorganisms.

In this experiment we will demonstrate the action of common bacteriostatic and bactericidal agents by use of the phenol coefficient method and the agar plate test. You and your partner will perform the experiments as a unit.

PERIOD I

PART A. PHENOL COEFFICIENT METHOD

MATERIALS

1. 2%, 1%, 0.5% solutions of phenol or Roccal (1% or 2%); 2% (by volume) solution of commercial Lysol.

2. Broth cultures (20 to 24 hr) of *Escherichia coli* (1 ml) and *Bacillus cereus* (1 ml).

3. 6 tubes of lactose broth with BTB indicator and Durham insert (for subculturing *E. coli*).

4. 6 tubes of nutrient broth (for subculturing *B. cereus*).

PROCEDURE

1. Each pair of students will work with one chemical in a certain dilution provided.

2. Label sterile lactose broth as follows: 1 tube as untreated control and remaining tubes as E.c. 1, 5, 10, 15, and 30 min. Label sterile nutrient tubes as above for subculturing *B. cereus.*

3. Inoculate 1 loopful of *E. coli* into a tube of lactose broth labeled as control. Do the same for *B. cereus* using nutrient broth.

4. a. Aseptically pipette 1.0 ml of your assigned chemical into a tube containing 1.0 ml suspension of *E. coli.* Note exact time and mix well.

   b. Aseptically pipette 1.0 ml of your assigned chemical into a tube containing 1.0 ml suspension of *B. cereus.* Note exact time and mix well. (Note: concentration of the chemical assigned to you is now reduced by 1/2. Why?)

   c. Incubate all tubes in a 20C water bath.

5. a. After 1, 5, 10, 15, and 30 minutes, subculture a loopful from the *E. coli*–chemical mixture to a tube of sterile lactose broth suitably labeled.

b. Repeat Step 5a for the *B. cereus*–chemical mixture, except inoculate into nutrient broth.

6. Incubate all the subcultured tubes in a 30C incubator until next period.

PART B. AGAR PLATE TEST

MATERIALS

1. Broth cultures (20 to 24 hr) of *Escherichia coli, Staphylococcus aureus,* and *Bacillus cereus.*
2. 3 sterile Petri dishes.
3. 1 bottle of melted nutrient agar (100 ml per bottle).
4. 3 tubes of nutrient agar, soft (0.7% agar).
5. Several sterile filter paper disks in a Petri dish.
6. Antimicrobial agents: tubes of 1% phenol, 70% ethyl alcohol, 1% HCl, 1:1000 diluted crystal violet, solution of soap, a mouth wash, etc.

PROCEDURE

1. Divide the bottom of 3 sterile Petri dishes into 6 equal sectors and place numerals 1 to 6 at the periphery of the dishes. Label the lid of the first dish as *E. coli,* that of the 2nd dish as *S. aureus,* and that of the last dish as *B. cereus.* (Note: one culture is gram-negative, one is gram-positive, and the third is a gram-positive sporeformer.)

2. Obtain one bottle of melted nutrient agar from the water bath. Wipe off all the water from the exterior of the bottle. Aseptically pour agar into the 3 labeled dishes. One bottle of agar should be enough to make 6 plates; give the remaining agar to your neighbor. Allow to solidify.

3. Obtain a tube of melted soft nutrient agar from the water bath (45 to 50C). Aseptically pipette 0.2 ml of *E. coli* into the agar. Mix well by rotating the tube between your palms and pour the entire contents into a properly labeled plate. Spread the contents evenly by tilting the plate back and forth. Allow to solidify. Do the same for *S. aureus* and *B. cereus.*

4. a. Sterilize your forceps (alcohol flame sterilization) and pick up one sterile filter paper disk from a Petri dish.

   b. Carefully saturate the disk with your assigned chemical. (Remove excess reagent by touching the disk to the interior of the tube.)

   c. Aseptically place the disk on the properly labeled sector and tap gently to insure closer contact with the surface of the agar.

5. Invert the plates and incubate in a 30C incubator until next period.

PERIOD II

PART A. PHENOL COEFFICIENT METHOD

PROCEDURE

1. Examine the growth in the subcultured tubes of *E. coli* and *B. cereus* by comparing with growth in untreated control. The growth of *E. coli* is readily detected by a change in BTB indicator (green changed to yellow) and by gas trapped in the Durham insert. The growth of *B. cereus* in nutrient broth is indicated by turbidity of the broth caused by cell multiplication. If the broth is clear, tap the bottom of the tube to make sure no sediment (settled cell mass) is present. Growth means that the chemical has not killed the bacteria in the time allowed for contact.

2. If the broth is clear throughout and no sediment is present, it may be concluded that the chemical has killed the bacteria in the contact time.

3. Record all your observations as well as those of your neighbor by exchanging your data.

PART B. AGAR PLATE TEST

PROCEDURE

1. Examine the inhibitory action of chemicals on each of the 3 cultures used. The clear zone surrounding the disks indicates bacteriostatic or bactericidal action of the chemical. If the growth of the organism occurred at the edge of the disk, it indicates that the substance had no effect on the culture.

2. Measure the diameter of the zone of inhibition in mm and record your readings in the space provided.

0.1 N NaOH

5 ml milk
+ 5 ml water
in flask
30 drops

.8 ml

5.0 ml milk
5.0 ml H_2O

5 drop

.4 ml
2.4 ml

2.0 ml.

6 ml.

used tap H_2O

2 day old

5.0 ml milk
5.2 ml water

NaOH 0.1
0.0
3.6 ml
used
3.6

1 day old

5.0 ml milk
5.0 ml water

3.6 ml
7.4
3.5

RESULTS

PART A. PHENOL COEFFICIENT METHOD

1. Record your observations from the phenol coefficient method in the table below. Use + to indicate growth and − for no growth.

Final Concentration of Chemicals	Cultures											
	E. coli						*B. cereus*					
	Cont.	*Exposure Time in Min*					Cont.	*Exposure Time in Min*				
	0	1	5	10	15	30	0	1	5	10	15	30
1%phenol												
0.5% phenol												
0.25% phenol												
1%Roccal												
0.5% Roccal												
1% Lysol												

PART B. AGAR PLATE TEST

1. Record the diameter of the zone of inhibition in mm.

Concentration and Chemical Assigned	Cultures		
	E. coli	*S. aureus*	*B. cereus*
1.			
2.			
3.			
4.			
5.			
6.			
7.			
8.			

DISCUSSION

1. In the phenol coefficient method, what is the reason for using the specified test organisms?

2. Does the agar plate test method necessarily measure the killing effectiveness of a chemical? Why?

SECTION

IV

EXPERIMENTS IN BACTERIAL GENETICS

EXPERIMENT 21

Variability in Pigment Formation

Although many bacterial species do not produce pigment, a fair number do. In some chromogenic species the pigment is confined within the cell and thus a mass of bacterial cells, such as the colony, appears colored. In other species, pigment is excreted into the environment and colors the medium. Pigment production is a striking cultural characteristic and the property is of value in the identification and classification of bacteria when they are studied under defined conditions.

Genotype is defined as the entire array of genes possessed by the cell, that is, the cell's genetic constitution; *phenotype* is the genetic constitution as expressed in a given environment, that is, the array of morphologic and physiologic properties observed under a given set of conditions. In some species of bacteria, chromogenesis is a phenotypic manifestation of the genotype. In other words, the ability to produce pigment and the intensity of the pigment produced are characteristics which are genetically controlled, but they may or may not be expressed, depending on such environmental conditions as the composition of the medium and the conditions of incubation. The genotype represents the total potential capability of the cell, whereas the phenotype represents those characteristics actually expressed.

Examples of common bacterial species which produce and retain pigment intracellularly are: *Serratia marcescens,* red pigment; *Sarcina lutea,* lemon yellow pigment; *Staphylococcus aureus,* golden yellow pigment; and *Chromobacterium violaceum,* violet pigment. Some bacterial species which produce and excrete pigments into the medium are: *Pseudomonas fluorescens,* green to yellow-green pigment; and *Pseudomonas aeruginosa,* which may produce red (pyorubin), blue-green (pyocyanin), or brown (pyoverdin) pigment, depending on the particular strain.

Some pigments, such as that from *P. fluorescens,* are water soluble and fluoresce when viewed with ultraviolet light. Others, such as prodigiosin (red pigment produced by *S. marcescens*) and pyocyanin are water insoluble. Some pigments (such as prodigiosin and pyocyanin) appear to have antimicrobial activities. However, the role of pigment production by bacteria in a microbial ecosystem is not clearly understood.

In this experiment we will demonstrate the effect of environmental conditions, both media composition and incubation temperature, on pigmentation. The experiment will be performed by you and your partner.

MATERIALS

1. Cultures of *Pseudomonas fluorescens, P. aeruginosa* strain R (strain which produces pyorubin) and strain B (strain which produces pyocyanin), *Serratia marcescens,* and *Micrococcus flavus* (grown in nutrient broth). Note: your instructor will also provide plate cultures of these organisms so that during Period II you may observe for demonstration of fluorescent pigment.

2. 10 sterile nutrient agar slants.

3. 10 slants of King's Medium A (peptone, 20.0 g; glycerol, 10.0 g; K_2SO_4, 10.0 g; $MgCL_2$, 1.4 g; agar, 15.0 g; in 1000 ml distilled water).

PROCEDURE

PERIOD I

1. Label 1 nutrient agar slant as P.f. 30C and 1 as P.f. 37C. Similarly label 2 agar slants for each of the remaining cultures.

2. Do the same with King's Medium A.

3. Aseptically inoculate each pair of nutrient agar slants and King's Medium A agar slants with one loopful of the labeled pure culture per slant.

4. For each culture, incubate one nutrient agar slant and one King's Medium A slant (labeled as 30C) in an incubator set at 30C. Incubate the second set of slants (labeled as 37C) in an incubator set at 37C.

PERIOD II

1. Examine each agar slant for presence or absence of pigment and record your observations in the space provided.

2. Examine the pigment color produced by each culture. Pigment color may vary with different incubation temperatures as well as with different media on which the cultures are grown. Record your observations in the space provided.

3. Determine whether the pigment is intracellular or excreted into the agar slants. If it is intracellular, no pigment will diffuse into the agar medium. Record your observations in the space provided.

4. Observe the demonstration plates inoculated with each of the five cultures under UV light. Under UV light, the pigment produced by *P. fluorescens* displays an eerie, blue-green iridescent effect. Observe the remaining cultures to determine whether the organisms produce fluorescent pigment. If

you are uncertain, consult your instructor. Record your results in the space provided.

5. Extraction of pyocyanin and pyorubin: Your instructor will extract pyocyanin and pyorubin from the slants inoculated with *P. aeruginosa* strains R and B. Record your observations in the space provided.

inoculate

(8) 2 Na – P.f., P.a. (CB), S.M., S.l.

(8) 2 King – " " " "

EXPERIMENT 22

Recombination by Conjugation

The primary process of reproduction by procaryotic organisms such as bacteria and some of the eucaryotic fungi (Class Deuteromycetes or Fungi Imperfecti) is asexual, that is, by either fission or budding. The process of nuclear fission in procaryons may seem haphazard (when compared with the reproductive processes of higher plant and animal cells), but the replication and division of DNA in procaryotic cells is nearly as exact and effective as the hereditary mechanisms of eucaryotic organisms.

Some procaryotic organisms may also carry out a type of sexual reproduction, but not true sexuality as observed in eucaryotic organisms—not a combining of genetic material from two germ cells to form a new individual. In the sexual reproduction of bacteria, transfer of genetic material is always *unidirectional:* genetic material is always transferred from a donor or + strain to a recipient or − strain, and the genes transferred are usually only a fraction of the total genetic capability of the donor strain (see Fig. 22-1). Nevertheless, the result is that the recipient acquires some different genetic material and may now express characteristics attributable solely to the donor.

There are at least four mechanisms of gene transfer observed in bacteria: *transformation, conjugation, transduction,* and *F-duction* (students are expected to understand the definitions of all these terms). See text for further details.

In this experiment we will demonstrate one of these processes: conjugation. The process involves the transfer of genetic material by an actual union or mating of bacteria. Mating pairs of bacteria have been observed microscopically. (See the references on the demonstration table.) The mating bacteria are sexually differentiated into *donor* (+) and *recipient* (−) types. The genetic material is transferred through some sort of bridge, possibly the F pili of the donor (see text); the donor cell transfers part of its genome to the recipient cell. The amount of DNA transferred will depend on whether the conjugal bridge is severed before transfer is completed (usually within 30 minutes; see Fig. 22-1). Genome transfer follows a definite sequence: always B after A, and so forth. If recombination then takes place in the recipient, it becomes a new individual having some characteristics of the donor. For example, the recipient may receive

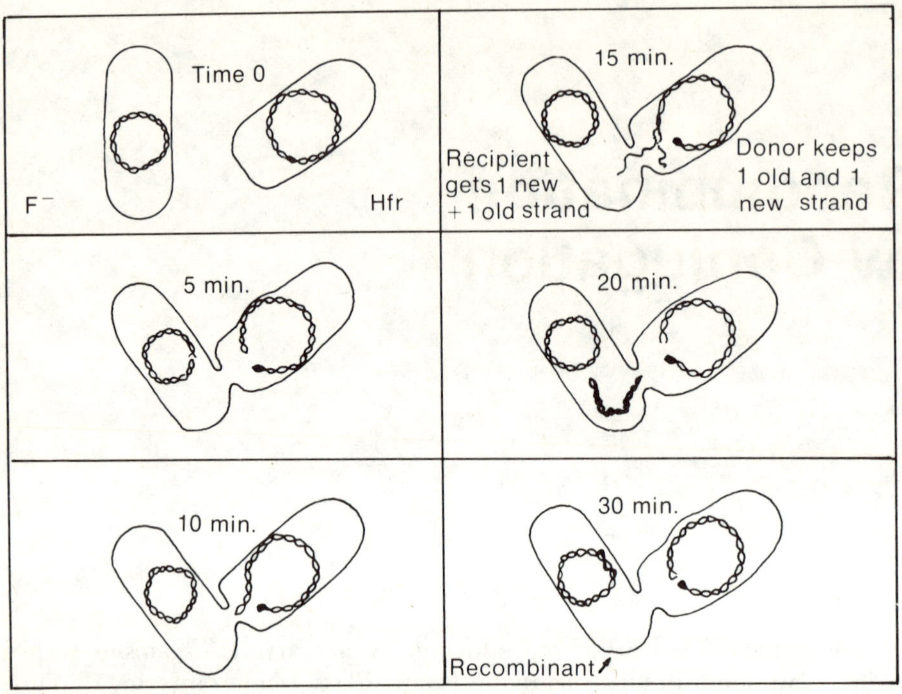

FIGURE 22–1. Possible mechanism of DNA transfer during conjugation.

a gene responsible for the synthesis of a certain amino acid which it previously lacked.

If we are to detect new progeny resulting from conjugation, mating types must possess special characteristic(s), called *genetic markers,* which can be easily recognized. The markers are usually obtained by mutating the *prototrophs* (wild type or parent type) through mutagenic agents such as ultraviolet irradiation or mitomycin C. The mutants deficient in some physiological attribute are now called *auxotrophs.* A marker can be any of a number of traits, such as antibiotic sensitivity or nutritional deficiency. A nutritional marker such as thr^- met^+ means that among the amino acids required for growth, the strain of bacteria can synthesize methionine but not threonine from the minimal medium containing only glucose, ammonia, and inorganic salts. If this organism is to grow and reproduce, then the medium must be supplemented with preformed threonine, or the organism must first receive a gene, thr^+, from donor cells. A strain identified as thr^+ met^- is similarly unable to grow in the minimal medium. However, when these two nutritionally marked stains are mixed and genetic recombination occurs following conjugation, some of the recombinants are able to grow upon the minimal medium because they now possess both genes (thr^+ met^+). The progeny thus demonstrate the same genetic potential as the prototrophs.

This experiment will be performed by you and your partner.

PERIOD I

MATERIALS (work in pairs)

1. 1 broth culture of *Escherichia coli* (10^8 cells/ml), auxotroph nutritionally marked (*thr*$^+$ *met*$^-$). The culture will serve as donor or + and is labeled M (donor or + strains are sometimes referred to as "male").

2. 1 broth culture of *E. coli* (10^8 cells/ml), nutritionally marked (*thr*$^-$ *met*$^+$). The culture will serve as recipient or − and is labeled F (recipient or − strains are sometimes referred to as "female").

3. 2 sterile tryptone–yeast extract–glucose agar (TYEG) plates (see Appendix for formula).

4. 5 sterile minimal medium (MM) agar plates (see Appendix for formula).

5. 1 sterile, empty, metal-capped test tube.

6. 5 sterile 9.0 ml distilled water dilution blanks.

7. 1 Petri dish turntable.

8. 1 50-ml beaker with 95% ethyl alcohol.

9. 1 bent glass spreading rod.

10. 1 canister of pipettes.

PROCEDURE

1. Label 1 each TYEG and MM agar plates as (M 1:10) and 1 each TYEG and MM plates as (F 1:10).

2. Label 1 sterile empty tube as (M × F). This tube is used for conjugation.

3. a. Aseptically pipette 1.0 ml of M culture into a 9.0 ml sterile water blank and mix thoroughly. Note the donor culture is now diluted to 1:10. Label the tube (M 1:10).

   b. Using the diluted (M 1:10) donor culture, prepare a streak culture on TYEG agar plate labeled (M 1:10). Do the same for the MM agar plate labeled (M 1:10). We anticipate TYEG agar will support growth of both M and F cultures, since the medium contains all the preformed amino acids including methionine and threonine, but MM agar should not support growth of either culture.

   c. Repeat Steps 3a and 3b using the culture labeled F.

4. a. Aseptically pipette 1.0 ml of (M) suspension into the conjugation tube labeled (M × F).

b. Aseptically pipette 1.0 ml of (F) suspension into the conjugation tube labeled (M × F). *Note the time* of addition of culture F to culture M in the conjugation tube. Shake the (M × F) mixture thoroughly and allow the cultures to conjugate for 30 minutes.

5. a. After the conjugation of (M × F) has proceeded for 30 minutes, pipette aseptically 1.0 ml of this mixture into a sterile 9.0 ml water blank. Mix the contents of the tube thoroughly and label the tube (M × F 1:10). See Figure 22-2.

b. Aseptically pipette 1.0 ml (M × F 1:10) suspension into a second sterile 9.0 ml water blank. Mix thoroughly and label the tube (M × F 1:100).

c. Similarly pipette 1.0 ml (M × F 1:100) suspension into a third sterile 9.0 ml water blank. Mix thoroughly and label the tube (M × F 1:1000).

6. a. Aseptically transfer 0.1 ml of (M × F 1:10) dilution to the surface of one of the MM agar plates.

b. Place the plate on a turntable. Sterilize the glass spreading rod by dipping in alcohol and passing quickly through a flame (see Fig. 17-1). Then spread the suspension evenly over the entire surface of the agar. Label the plate (M × F 1:100).

7. Do the same for (M × F 1:100) and (M × F 1:1000) by repeating Steps 6a and 6b. Label the plates (M × F 1:1000) and (M × F 1:10,000). Note: the final dilution has increased by a factor of 10 for each of the 3 dilutions spread. This is because only 0.1 ml of each dilution was used.

8. Allow the spread plates to stand at room temperature for 5 minutes, then incubate the plates, inverted, for 2 to 4 days at 30C.

PERIOD II

PROCEDURE

1. Examine the streak plates of the donor (*thr*$^+$ *met*$^-$) and recipient (*thr*$^-$ *met*$^+$) cultures on TYEG and MM agar media. On which medium was each strain able to grow and form colonies? Record your observations in the space provided.

2. Count the number of colonies on the MM agar plates labeled (M × F 1:100, 1:1000, and 1:10,000). Select the plate with colony numbers between 30 and 300 and multiply the colony number by the dilution on the plate. This will give the recombinant colony-forming units per ml of the (M × F) mixture. Record your observations in question 2 under Results.

3. Calculate the percentage of recombinants in the (M × F) mixture. Remember the (M × F) mixture contained approximately 1×10^8 cells/ml.

$$\% \text{ recombinants} = \frac{\text{number of recombinant colony-forming units/ml} \times 100}{\text{number of cells/ml in (M} \times \text{F) mixture}}$$

Record your results in question 3 of Results.

RESULTS

1. Fill in the table below with your observations.

	Growth on TYEG Agar	Growth on MM Agar
(*Thr*$^+$ *met*$^-$) donor		
(*Thr*$^-$ *met*$^+$) recipient		

2. Record the number of colonies formed on MM agar.

	10^{-2} Plate	10^{-3} Plate	10^{-4} Plate
(M × F)			

Therefore, recombinant colony-forming units/ml of (M × F) mixture is ________.

3. The percentage of recombinants appearing as recombinant colony-forming units/ml of the (M × F) mixture is ______________.

DISCUSSION

1. What would you anticipate if the contents of the conjugation tube were blended after 5 minutes, immediately diluted, and the sample plated on MM agar? Would you anticipate more colony-forming units or fewer colony-forming units?

2. Can you envision any circumstances in which colonies might appear on the MM agar plates even though recombination had not taken place?

Isolation of Bacterial Mutants

Bacterial mutants may arise spontaneously, or they may be produced through the use of mutagenic agents *(mutagens)* such as ultraviolet rays (UV), X-rays, nitrous acid, alkylating agents (nitrogen mustard and mitomycin), and base analogues. The ways in which various mutagens alter DNA are beyond the scope of this introduction, but the essential point is that the mutant has an altered genetic code; hence, the mutant often expresses a phenotype clearly different from that of the parent. The difference, however, depends largely upon how severely the genetic code was rearranged. A mutation may involve so trivial an alteration as to produce no easily recognized changes in the progeny; or, the alteration may be so severe as to be lethal (the mutant dies and no progeny are produced).

Some of the categories of mutations frequently encountered by microbiologists are:

Resistance Mutants resistant to inhibitory agents such as antibiotics, bacteriocins, etc.

Biochemical Auxotrophic, fermentation, and pigmentation mutants.

Structural Mutants with altered cellular morphology: lack flagella, pili, proper cell wall, etc.

Colonial Mutants with alterations which are apparent by changes in colonial morphology, e.g., rough-type (R) mutants from smooth-type (S) culture.

Antigenic Mutants with altered intracellular or surface antigens; may or may not coincide with altered colonial morphology.

Virulence Mutants with higher or lower virulence than parental type. May or may not coincide with changes in colonial morphology, antigenicity, biochemistry, etc.

While these kinds of mutations take place spontaneously, they do so at very low rates and are generally not recognized unless some sort of selective environment favoring the growth of the mutant is present. Mutagens, which can increase mutation rates 10 to 100,000 fold, together with specially developed selection techniques, are generally employed to isolate mutants of the desired type. Not only has the production of valuable industrial products been greatly increased through the use of mutants, but much of modern microbial genetics and molecular biology has also depended on the availability of certain types of mutants.

In this experiment we will concentrate on the isolation of antibiotic-resistant mutants. The work will be complicated by the fact that we may encounter either one-step or multi-step resistant mutants, depending on the culture and antibiotic chosen for this study. For example, if a low level of penicillin is added to a broth heavily inoculated with a penicillin-susceptible strain of bacteria, a few of the bacteria will be resistant mutants and will grow in the presence of the penicillin. If their progeny are subsequently grown in the presence of somewhat higher amounts of penicillin, a few mutants will survive and grow. If this process is continued, using successively higher concentrations of penicillin, mutants with resistance to a high concentration can be isolated. No resistant mutants will normally be obtained, however, if the susceptible culture is exposed to a high concentration of penicillin in the first step.

With streptomycin, a different situation may be encountered. In a large population of bacteria, there may be mutants which are resistant to either low, medium, or high concentrations of streptomycin. Thus, highly resistant mutants can often be obtained in one step. Other examples of one-step mutants would be phage-resistant and colicin-resistant mutants.

MATERIALS

1. 4 deep (10 ml) tubes of nutrient agar (supplemented with 0.2% yeast extract and 2% glycerol).
2. 2 sterile Petri dishes
3. 1 tube of heavy *Escherichia coli* B suspension.
4. 1 tube of sterile streptomycin solution.*
5. 1 tube of sterile penicillin G solution.*
6. 1 canister of pipettes.
7. 1 bent glass spreading rod.
8. 1 50-ml beaker with 95% alcohol.
9. 1 plastic bag sufficient to hold two plates.

PROCEDURE

PERIOD I

1. Pour a melted, cooled (45 to 50C) tube of nutrient agar into a tilted sterile Petri dish as illustrated in Figure 23-1a. Use aseptic technique and be sure the agar covers the entire bottom of the dish. Allow the agar to solidify

*The concentrations of streptomycin and penicillin to be used in these experiments will have been worked out in a preliminary experiment by your instructor. Generally, final concentrations in the upper agar layer of 200 U/ml of penicillin or 30 μg/ml of streptomycin are satisfactory. As an option, your instructor may provide some students with other types of inhibitors, adjusted to give an appropriate concentration in the upper agar layer. Suggested are: tetracycline (15 μg/ml), chloramphenicol (10 μg/ml), or sodium chloride (12%).

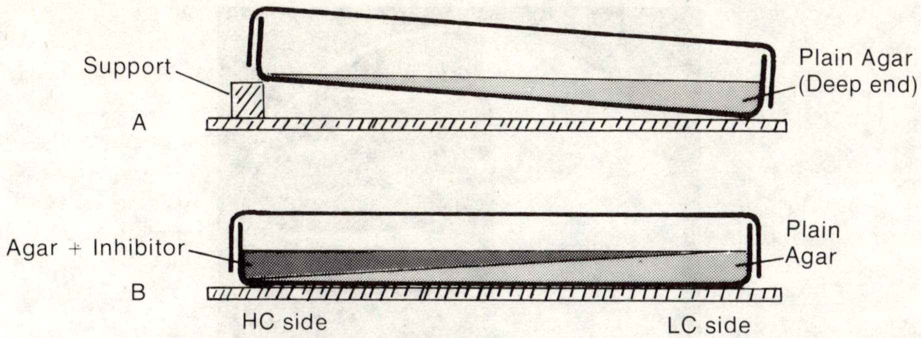

FIGURE 23–1. Preparation of a gradient plate.

completely before moving. Mark the deep end (see Fig. 23-1A) of the tilt plate LC (low concentration).

2. Into another melted, cooled (45 to 50C) tube of nutrient agar, pipette the appropriate amount of streptomycin solution as indicated by your instructor. Mix thoroughly.

3. Place the plate from Step 1 on a flat surface. Pour the nutrient agar containing the antibiotic into the plate and allow to solidify (Fig. 23-1B). During subsequent incubation, the downward diffusion of the antibiotic will establish a uniform concentration gradient proportional to the thickness ratio of the two agar layers.

4. In a similar fashion prepare a gradient plate using the penicillin solution.

5. Aseptically pipette 0.1 ml of the heavy *E. coli* B suspension onto each of the two gradient plates.

6. Sterilize the spreading rod by dipping in alcohol and quickly passing through the flame. Spread the culture evenly over the surface of one plate. Sterilize the rod again and spread the inoculum over the second plate.

7. Invert plates, insert in a bag, and incubate at 30C for 4 to 7 days. Plastic bags help to prevent the agar from drying out.

PROCEDURE

PERIOD II

1. Examine the antibiotic gradient plates and record your observations. Remember that the side of the plate marked LC has the lowest concentration of antibiotic.

2. On the penicillin plate, select a colony beyond the area of confluent growth (i.e., a colony of bacteria believed to have some resistance). With a sterile loop, streak this colony across the surface toward the high concentration (HC) side of the gradient plate.

3. Do the same for the plate containing streptomycin or one of the optional inhibitors.

4. Incubate the plates as before for 2 to 4 days.

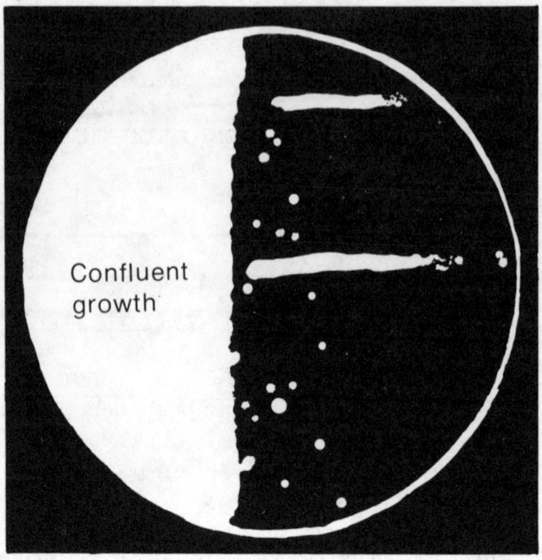

FIGURE 23–2. Resistant colonies are seen just beyond the area of confluent growth on this gradient plate. Two resistant colonies were streaked toward the HC side of the plate and the plate was incubated for a second time. This resulted in the isolation of second-step mutants on the far (HC) side. (From Szybalski and Bryson, *J. Bacteriol.* *64*:489,1952.)

PROCEDURE

PERIOD III

1. Examine both plates and record your observations. Compare results with others in the laboratory. Some students should obtain penicillin gradient plates demonstrating the growth pattern shown in Figure 23-2. What is the meaning of the isolated colonies on the HC side of the plate? Did the streptomycin plate give a similar or different pattern of growth?

RESULTS

1. Fill in the table below with your observations from Period I.

	Penicillin Plate	Streptomycin Plate
Isolated colonies present in central area of plate		
Isolated colonies on HC side of gradient		

2. Did you obtain evidence of one-step mutation for streptomycin? For penicillin?

3. In the last period, did you see evidence of a two-step increase in penicillin resistance? Explain.

DISCUSSION

1. Can you think of any practical application of this technique?

2. What might be the consequences of antibiotic resistance in the treatment of infectious disease with antibiotics?

3. How might antibiotic resistance be avoided?

4. Do you recall from lecture the mechanisms involved in antibiotic resistance? What is penicillinase?

SECTION

V

ASPECTS OF MICROBIAL ECOLOGY

EXPERIMENT 24

Bacterial Commensalism

Microbial ecology is the study of the relationship between microorganisms and their environment. The environment may be soil, rhizosphere, water, rumen, oral cavity, even body surfaces, to mention a few. Each of these environments has a microbial population which differs from that of others qualitatively, quantitatively, and with regard to microbial interactions. The types of interactions observed are: *symbiosis* (ecto- or endosymbiosis, Experiment 26); *antagonism* (Experiment 18); *neutralism* (relationship between two or more organisms with neither kind of organism affecting the existence of the other); *commensalism* (Experiment 24); and *synergism* (Experiment 25). Microbial interactions are extremely important in the everyday life of all microorganisms as they rarely exist alone in nature and often have to rely, in part, on other species for survival.

We frequently encounter the relationship in which one organism benefits and the other is unaffected; this phenomenon can be considered a form of *commensalism*. In commensalism the microorganism which benefits by the association is termed the *commensal organism,* and the organism unaffected by the relationship is said to be the *host.* The host confers its beneficial effects through its metabolic activities, which alter the physicochemical environment and favor the growth of the commensal organism. How does a commensal differ from a parasite?

There are many examples of commensal relationships in the world of microbes. For instance, strictly anaerobic organisms may be able to survive in an environment exposed to free oxygen if they are grown together with facultative organisms which deplete the free oxygen, thus making the environment favorable for anaerobic growth. (It should be remembered that since strict anaerobes will grow only in the absence of free oxygen, vegetative cells of strict anaerobes may be "poisoned" or killed when exposed to high oxygen tension. Facultative organisms, on the other hand, are able to grow with or without free oxygen.) Similarly, certain organisms not only synthesize growth factors for themselves but release some of these substances into the environment on which commensal organisms depend for growth. For example, *Staphylococcus aureus* produces V factor (presumably NAD or NADP), which supports the growth of *Haemophilus influenzae;* certain yeast cells of bacteria able to grow in extremely osmophilic environments may reduce the osmolarity of the solution by producing metabolic water, thus allowing the growth of other microorganisms normally incapable of growing in such an environment; an organism unable to grow in á starch medium may be able to grow if it is cultured with an organism which produces extracellular amylase to hydrolyze the starch to maltose.

In a natural ecosystem, most of the transformations of sulfur, carbon, and nitrogen occur as the result of microbial commensalism. For example, nitrogen-fixing organisms convert atmospheric nitrogen to cellular nitrogen; cellular nitrogen is eventually converted to ammonia through ammonification; ammonia is oxidized to nitrite and nitrate, which is used by higher plants, reduced to nitrite, or denitrified to gaseous nitrogen by certain bacteria. (For a detailed discussion of the nitrogen cycle, see Experiment 26.)

The objective of this experiment is to demonstrate two simple types of commensalism widely found in nature.

PERIOD I

PART A. NUTRITIONAL COMMENSALISM (THE EXTRACELLULAR HYDROLYTIC ENZYME ACTION OF ONE ORGANISM ON A SUBSTRATE RESULTING IN THE SUPPORT OF GROWTH OF ANOTHER SPECIES WITHOUT AFFECTING THE GROWTH OF THE HOST)

MATERIALS (work in pairs)

1. 3 plates of casein agar (sodium caseinate, 2.0 g; glucose, 1.0 g; $MgSO_4$, 0.2g; K_2HPO_4, 0.2 g; $FeSO_4$, trace; agar, 15 g; in 1000 ml distilled water).
2. Broth cultures of *Bacillus cereus* and *Pseudomonas fluorescens.*

PROCEDURE

1. Label 3 casein agar plates as follows: first as B.c. control, second as P.f. control, and third as B.c. + P.f.
2. Prepare a streak plate of *B. cereus* on the plate labeled B.c. control. Do the same for *P. fluorescens.*
3. Inoculate the plate labeled B.c. + P.f. as follows:
   a. Make 2 steaks across the plate approximately 3 cm apart using a loopful of *B. cereus.*
   b. Streak the uninoculated area between the 2 lines of *B. cereus* with *P. fluorescens* without touching the *B. cereus* streaks.
4. Incubate all of the plates at 30C to 2 to 4 days.

PART B. PHYSICAL COMMENSALISM (COMMENSAL ASSOCIATION ALLOWING THE GROWTH OF FACULTATIVE AND STRICTLY ANAEROBIC SPECIES IN AN ENVIRONMENT EXPOSED TO FREE OXYGEN)

MATERIALS (work in pairs)

1. Culture of *Staphylococcus aureus* in nutrient broth; culture of *Clostridium sporogenes* (strict anaerobe) in fluid thioglycollate broth.

2. 3 tubes of nutrient broth.

3. 2 tubes of fluid thioglycollate broth.

PROCEDURE

1. Label 1 tube of fluid thioglycollate broth as S.a. and the other as C.s. Label 1 tube of nutrient broth as S.a., the 2nd tube as C.s., and the 3rd tube as S.a. + C.s.

2. Aseptically inoculate the tube of fluid thioglycollate broth labeled C.s. with a loopful of *C. sporogenes* suspension from the bottom of the culture tube. Similarly inoculate the tube of nutrient broth labeled C.s.

3. Repeat Step 2 using the culture of *S. aureus.*

4. Aseptically inoculate the tube of nutrient broth labeled S.a. + C.s. with a loopful each of *S. aureus* and *C. sporogenes.*

5. Incubate all the tubes at 37C for 24 to 48 hr.

PERIOD II

PART A. NUTRITIONAL COMMENSALISM

MATERIALS

1. 1% HCl.

PROCEDURE

1. Observe the growth of *B. cereus* and *P. fluorescens* on the control plates. *B. cereus* should produce extracellular caseinase which hydrolyzes casein to peptides. *P. fluorescens* is unable to produce the enzyme, thus shows little or no growth on the medium. Make a note of the amount of growth, the size and pigmentation of the growth, and the size of the clear, hydrolyzed zone produced by *B. cereus.* After you have observed the growth characteristics, flood the plates with 1% HCl. The clear zone indicates areas of casein hydrolysis; the white, opaque precipitate indicates no hydrolysis.

2. Carefully observe the plates labeled B.c. + P.f. If there is growth of *P. fluorescens* it should be near the *B. cereus* streaks or in the clear zone produced by *B. cereus*. *Pseudomonas* colonies usually appear iridescent and flat.

3. Make a Gram stain of cells from either of the 2 streaks *(B. cereus)* and from the iridescent colonies *(P. fluorescens)* to ascertain the commensal relationship of these two organisms. Record all of your observations in the space provided.

PART B. PHYSICAL COMMENSALISM

PROCEDURE

1. Observe the growth of *S. aureus* and *C. sporogenes* in fluid thioglycollate broth. Note and compare the regions in which each of these two cultures grew. Remember, *S. aureus* is facultative and *C. sporogenes* is a strictly anaerobic sporeforming organism. Record your observations in the space provided.

2. Observe the growth in the tubes of nutrient broth. Was there any growth in the tube labeled S.a. + C.s.? Record your observations in the space provided.

3. Make a slide smear by obtaining a loopful of cell suspension from the bottom of the nutrient broth tube labeled S.a. + C.s. Air dry, heat fix, and stain with Gram's stain. As these organisms are in commensal relationship, you should be able to observe rod-shaped cells with a swollen end *(C. sporogenes)* as well as spherical or coccoid cells *(S. aureus)*. Sketch a representative field in the space provided.

RESULTS

PART A. NUTRITIONAL COMMENSALISM

1. Fill in the following figures with your observations.

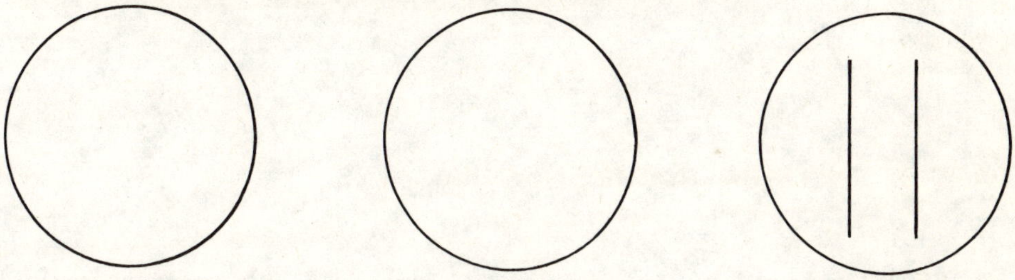

Control B.c. Control P.f. B.c. + P.f. (two lines indicate *B. cereus* streaks)

PART B. PHYSICAL COMMENSALISM

1. Fill in the following figures with your observations.

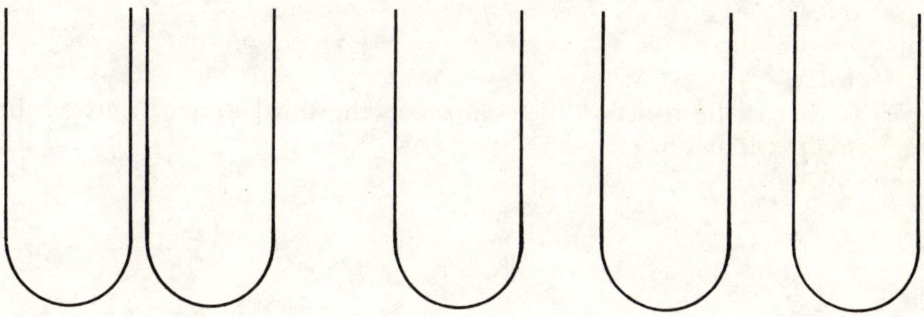

S.a. C.s.
Fluid thioglycollate broth

S.a C.s. S.a. + C.s.
Nutrient broth

DISCUSSION

1. Discuss why *P. fluorescens* was able to grow in the clear zone produced by *B. cereus.*

2. Why were both *S. aureus* and *C. sporogenes* able to grow in fluid thioglycollate broth but *C. sporogenes* was unable to grow in nutrient broth?

3. What is the function of *S. aureus* in the third tube of nutrient broth (S.a. + C.s.)? Discuss briefly.

EXPERIMENT 25

Bacterial Synergism

In the "natural environment" there are many situations in which two dissimilar organisms benefit from each other's existence; this symbiotic relationship is frequently referred to as *synergism* or *mutualism.* A classic example of bacterial-plant synergism will be discussed in the next experiment under symbiotic nitrogen fixation. In bacterial synergism, one frequently observes a characteristic interaction between two organisms which results in an effect or metabolic product neither organism can produce alone. Unfortunately, many of the examples that could be cited are not easily demonstrated in an elementary course. It is known, for instance, that certain strains of *Staphylococcus aureus* and *Proteus vulgaris* produce mortality when they are injected into mice concurrently, whereas equal numbers of either strain injected individually do not. Swine influenza is caused by the combined action of a bacterium *(Haemophilus suis)* and a virus *(Tarpesia suis).* Mixtures of different bacteria, such as those found in dairy starter cultures, are frequently more active in acid production than is any single strain in pure culture. Numerous examples of bilateral exchange of growth factors could be cited; for instance, biotin excreted by *P. vulgaris* aids the growth of *Bacillus polymyxa,* which in turn excretes nicotinic acid and aids the growth of *P. vulgaris.*

Another example of bacterial synergism is the production of gas from lactose by the combined action of *Proteus vulgaris* and *Staphylococcus aureus.* Both organisms are able to ferment lactose but individually neither produces gas from lactose fermentation. It has also been found that neither *Escherichia coli* nor *Streptococcus faecalis* is able to produce putrescine from arginine, but the combined action of these two organisms will produce putrescine. (Individually, *E. coli* is able to decarboxylate arginine to agmatine and *S. faecalis* is capable of converting arginine to ornithine. However, when these two cultures are grown together, *E. coli* is able to decarboxylate ornithine produced by *S. faecalis* to putrescine). The two types of synergism just mentioned will be demonstrated in the following experiment.

PERIOD I

PART A. PUTRESCINE PRODUCTION

MATERIALS

1. Broth cultures of *Streptococcus faecalis* and *Escherichia coli.*

2. 3 tubes of arginine broth in screw cap tubes (Bacto-peptone [Difco], 1.0 g; NaCl, 5.0 g; K_2HPO_4, 0.3 g; glucose, 1.0 g; arginine, 10.0 g; in 1000 ml distilled water). 1 ml of 1.6% brom thymol blue may be added to the medium to detect change in pH; the final pH of the medium should be approximately 7.2.

PROCEDURE

1. Label 3 tubes of arginine broth as follows: 1 tube as S.f., the second tube as E.c., the third tube as S.f. + E.c.

2. Inoculate the tube labeled S.f. with a loopful of *S. faecalis* inserted deep into the tube. Do the same for tubes labeled E.c. and S.f. + E.c.

3. Incubate the tubes at 30C for 2 to 5 days.

PART B. GAS PRODUCTION BY BACTERIAL SYNERGISM

MATERIALS

1. Broth cultures of *Staphylococcus aureus* and *Proteus vulgaris.*

2. 3 tubes of lactose broth with Durham inserts (beef extract, 3.0 g; peptone, 5.0 g; lactose, 10.0 g; 1000 ml distilled water; 1.0 ml of 1.6% solution of brom thymol blue as pH indicator).

PROCEDURE

1. Label 3 lactose broth tubes as follows: 1 tube as S.a., the second tube as P.v., the third tube as S.a. + P.v.

2. Inoculate the lactose broth tube labeled S.a. with 1 loopful of *S. aureus* and the tube labeled P.v. with 1 loopful of *P. vulgaris.* Inoculate the tube labeled S.a. + P.v. with 1 loopful of *S. aureus* followed by 1 loopful of *P. vulgaris.* (Be sure to sterilize inoculating loop with flame before and after each inoculation.)

3. Mix the broth by rotating the tube between the palms of your hands. Incubate the tubes at 37C for 2 to 5 days.

PERIOD II

PART A. PUTRESCINE PRODUCTION

PROCEDURE

1. Examine each tube for growth of organisms and record your observations in the space provided.

2. Production of putrescine and other amine compounds in the tube labeled S.f. + E.c. may be detected by a change in the indicator, BTB, to dark blue. Compare the color with the control, i.e., with the tubes inoculated with *E. coli* alone and with *S. faecalis* alone. Record your observations in the space provided. Note: conversion of arginine to various amine compounds will change the pH from the acidic to the basic side. Your instructor will provide you with the final pH of the medium.

PART B. GAS PRODUCTION

PROCEDURE

1. Remove the lactose tubes from the incubator and examine them for acid production. If acid is produced, the BTB indicator changes to yellow. Examine for gas production, especially in the tube labeled S.a. + P.v. If gas is present, bubbles should be trapped in the Durham tube. Record your observations in the space provided. If there is no change in the indicator, record as NC (no change); if no gas is produced, record as NG.

RESULTS

PART A. PUTRESCINE PRODUCTION

1. Record your observations in the table below.

Culture	Growth + or −	Describe change in color of indicator
S. faecalis		
E. coli		
S. faecalis + *E. coli*		

PART B. GAS PRODUCTION

1. Record your observations in the table below.

Culture	Growth + or −	Acid + or −	Gas Production + or −
S. aureus			
P. vulgaris			
S. aureus + *P. vulgaris*			

DISCUSSION

1. Compare bacterial synergism and commensalism.

2. Diagram the metabolic pathway of arginine to putrescine by the synergistic action of *E. coli* and *S. faecalis.* You may consult your text.

3. Give a few examples of bacterial synergism other than those mentioned in the introduction. You may consult any textbook you wish.

EXPERIMENT 26

Microorganisms and the Cycling of Nitrogen

The nitrogen cycle, as it occurs in nature, comprises at least four distinct transformation stages. These are summarized in Figure 26-1. The first stage is *nonsymbiotic* and *symbiotic nitrogen fixation,* in which the gaseous nitrogen of air is biologically converted to protein and amino nitrogen; the second stage, aerobic decomposition of protein by a variety of microorganisms with subsequent release of ammonia, is termed *ammonification* (anaerobic decomposition of protein is frequently termed *putrefaction,* since the process releases various metabolic products which produce putrid odors); the third stage, *nitrification,* is biologic oxidation of ammonia to nitrite and subsequent oxidation to nitrate; the fourth, *denitrification,* is the biologic reduction of nitrate to gaseous nitrogen.

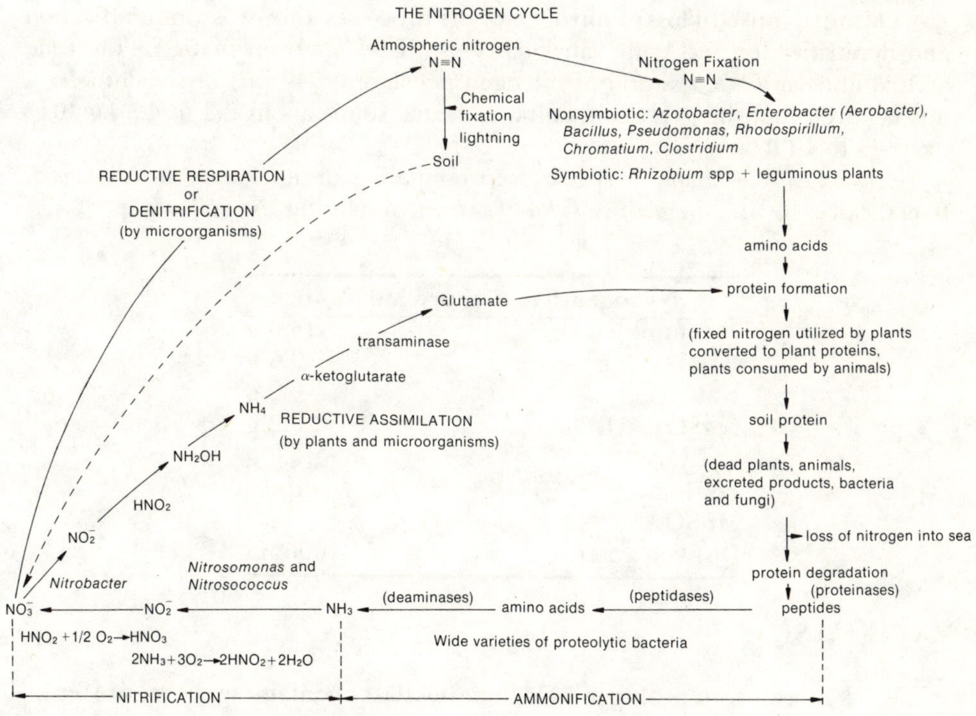

FIGURE 26–1. Generalized nitrogen cycle.

The purpose of the following experiments is to demonstrate four of the nitrogen transformation stages. Your instructor may assign all or only certain portions of the experiment, depending on the time available.

PART A. NONSYMBIOTIC AND SYMBIOTIC NITROGEN FIXATION

1. NONSYMBIOTIC NITROGEN FIXATION

Only a few bacteria and blue-green algae are capable of transforming molecular nitrogen into cellular (protein) nitrogen, which in turn eventually becomes available to other organisms. This process, conversion of gaseous nitrogen of the air into a "fixed form," is called *nitrogen fixation.*

Among the nonsymbiotic nitrogen fixers, some of the most thoroughly studied are species of the Genus *Azotobacter,* a group of free-living, aerobic soil organisms capable of utilizing atmospheric nitrogen as their only nitrogen source. Among anaerobic microorganisms, *Clostridium pasteurianum* is considered an important nonsymbiotic nitrogen fixer. Others, though their importance in nitrogen enrichment of soil is not well established, include some photosynthetic bacteria, such as *Rhodospirillum rubrum* and common gram-negative saprophytes of the Genus *Enterobacter (Aerobacter).*

The precise mechanism of nitrogen fixation is still not clear, but it appears to consist essentially of the reduction of free nitrogen to ammonia. The ammonia can then be coupled with α-ketoglutaric acid in the cell to form a very versatile amino acid, glutamic acid.

Owing to constant loss of nitrogen by the processes known as ammonification and denitrification, no truly satisfactory estimate has been made of the total annual nonsymbiotic fixation of nitrogen per acre (4047 m^2) of soil; however, it is believed that the average fixation in fertile soil is about 2.3 to 4.5 kg/4047 m^2/yr (5 to 10 lb/acre/yr).

Azotobacter vinelandii, a species representative of nonsymbiotic nitrogen-fixers, can grow in *nitrogen-free culture medium* containing the following:

Nitrogen-free Culture Medium	
Mannitol	15.0 g
K_2HPO_4	0.5 g
NaCl	0.2 g
$MgSO_4 \cdot 7H_2O$	0.2 g
$CaCO_3$	5.0 g
$FeCl_3$	trace
$MnSO_4$	trace
Distilled water	1000 ml

MATERIALS

1. 2 125 ml cotton-plugged Erlenmeyer flasks containing 40 ml of sterile nitrogen-free medium.

2. Pure culture of *A. vinelandii* grown in the above medium.

3. A sample of soil.

4. A canister of sterile pipettes.

PROCEDURE

PERIOD I

1. Inoculate one flask of the medium with 1.0 ml of the pure culture provided. Label the flask as A.v. and incubate on a rotary shaker (to promote aeration) set at approximately 160 rpm.

2. Inoculate the second flask with approximately 1 g of the soil provided. Label the flask and place in your desk cabinet. The reason for the static incubation will become clear later.

3. Prepare carbol-fuchsin stained smears of the pure culture of *A. vinelandii.* (Staining for 30 seconds will be sufficient.) Examine the slide under the oil immersion objective and describe and draw the cells in the space provided.

PERIOD II

1. After incubation for 5 days, observe the flask culture on the shaker for growth of *A. vinelandii.* Growth is evinced by increased turbidity over the settled $CaCO_3$. Compare with the control flask on the demonstration table. Examine the flask with soil (do not shake or disturb the medium) and describe its appearance. Are there any gas bubbles? Remove the cotton plug and take a sniff. Is there any particular odor? Gas bubbles and a rancid odor usually indicate active growth of *Clostridium pasteurianum.* If *Azotobacter* species (aerobic) are growing, a white to brownish film is usually formed on the surface of the medium.

2. Your partner will prepare smears from the soil flask as follows: Take one loopful of surface growth (from a filmy part) and spread it on the slide, left of center. Take one loopful of bottom growth (anaerobic) and spread it uniformly on the slide, to the right of center. Air dry, heat fix, and stain with carbol-fuchsin for 30 seconds. You will prepare hanging drop preparations of the surface and bottom growth. Spherical or ovoid cells occurring in pairs may be species of *Azotobacter,* and long, rod-shaped cells which appear to be bulged at the center or at one end may be *Clostridium* species. Record your observations in the space provided.

RESULTS

1. Make drawings of a few cells from each preparation. Observe the shape of individual cells and the characteristic arrangement. For soil preparations, note particularly the number and kinds of cells present.

Samples	Word Description of Broth Culture	Carbol-fuchsin Stained Smear	Hanging Drop Preparation
A. vinelandii 0 hr			
A. vinelandii 5 days			
Soil 5 days			

DISCUSSION

1. In hanging drop preparations of soil you have observed a variety of bacteria as well as protozoans. Why are non-nitrogen-fixing bacteria and protozoans able to multiply in the nitrogen-free medium?

2. *C. pasteurianum* is strictly anaerobic and *A. vinelandii* is strictly aerobic. Why are these two organisms capable of growing in the same flask? Refer to Experiment 24 if you are not certain.

2. SYMBIOTIC NITROGEN FIXATION

A mutually beneficial state in which two different forms of life grow together in close association is called symbiosis. Endosymbiotic relationships are found between leguminous plants (e.g., alfalfa, clover, beans, peas, and soy beans) and species of *Rhizobium.* These bacteria (rhizobia) are called root nodule bacteria because they enter the roots of legumes and produce nodules which contain millions of bacterial cells. The bacteria obtain all their necessary food material *except nitrogen* from the plant root and their nitrogen from the soil atmosphere; the fixed (combined) nitrogen of the rhizobia eventually becomes available to the plant. The presence of bacteria in the nodule thus improves plant growth, especially in a soil that contains little combined nitrogen. Since neither the plant nor the bacteria can fix nitrogen alone, we have an example of true *symbiotic nitrogen fixation.*

One of the unique characteristics of rhizobia is that when grown in root nodules they produce bizarre, irregular-shaped cells known as *bacteroids.* When transferred to suitable culture media, however, they appear as small, nonspore-forming, rod-shaped bacteria. Species of *Rhizobium* are differentiated and named primarily on the basis of the legumes they can "infect" and with which they are able to establish a symbiotic relationship: *Rhizobium trifolii* infects red, crimson, and white mammoth clover; *R. meliloti* infects alfalfa, sweet clover, and bur clover. Any group of legumes within which the root nodule bacteria are interchangeable is called a *bacterial plant* or *cross-inoculation group.* While soil usually contains a sufficient number of cross-inoculation groups, the current practice is to plant seeds which have been previously inoculated with a specific species of *Rhizobium* to insure nodule formation. This experiment is designed to isolate *Rhizobium* sp from a leguminous plant and to compare the growth of the rhizobia with the pure culture provided.

PERIOD I

MATERIALS (work in pairs)

1. A leguminous plant with root nodules.
2. Pure culture of *Rhizobium* sp on agar slant.
3. 3 sterile 10 ml water blanks.
4. 10 ml 70% ethyl alcohol in a test tube.
5. 2 plates of yeast lysate-mannitol agar.

Yeast Lysate–Mannitol Agar	
Mannitol	10.0 g
K_2HPO_4	0.5 g
$MgSO_4$	0.2 g
NaCl	0.1 g
Agar	15.0 g
Yeast water*	100.0 ml
Distilled water	900.0 ml

PROCEDURE

1. Each pair of students should obtain 1 leguminous plant with root nodules. Sketch a portion of the root with nodules in the space provided.

2. Each student should select several large nodules and remove the nodules with a piece of root attached. This is to facilitate washing of the nodules in the following steps.

3. Wash the nodules in a tube of sterile water by shaking for about 3 minutes.

*Yeast water is prepared by adding 100 g pressed, starch-free yeast to 1 liter distilled water. Heat for 3 hr in flowing steam, place in cotton-plugged flask, or sterilize in autoclave 90 min at 15 lb pressure, and allow to stand. Siphon off supernatant liquid, adjust to pH 6.6 to 6.8. (2 g/l of yeast extract may replace yeast water; however, the results obtained with yeast water appear to be superior to those obtained with yeast extract.)

4. Remove the nodules with flame-sterilized, cooled forceps by grasping a bit of the attached root.

5. Place the nodules in 70 per cent alcohol for 3 minutes.

6. Aseptically remove the nodules from the alcohol and rinse them in the second sterile water blank. (It is assumed that most of the bacteria present on the surface of the nodules were either washed off or killed by this treatment.)

7. Place the nodules on a clean slide and crush them by pressing another clean slide over them. Prepare a thin smear by moving the top slide over the lower one. Allow the lower slide to air dry. Place the upper slide on the workbench with the side in contact with the root nodule facing upward.

8. Using a flame-sterilized wire loop, transfer one loopful of water from the third sterile water blank onto the upper slide (side which was in contact with the root nodule) and mix the water with the specimen on the slide. Transfer a small amount of the mixture onto the agar plate and prepare a streak plate. Also prepare a streak plate from the pure culture provided. Label the plates; invert and incubate them in your desk cabinet for 2 to 5 days.

9. Heat fix the air-dried smear and stain with aqueous methylene blue for 3 minutes. Observe the slide with the oil immersion objective. Describe and draw a few bacteroids in the space provided.

PERIOD II

PROCEDURE

1. After incubation for 2 to 5 days, examine the streak plates prepared from the nodules and pure culture. Draw and describe typical colonies of *Rhizobium.*

2. Make methylene blue stained smears from one of the colonies studied and from the pure culture on the same slide in the usual manner.

3. Observe the stained preparations under the oil immersion objective and draw a few cells of each in the space provided.

RESULTS

1. Sketch a portion of the root with nodules attached.

2. Draw and describe bacteroids from the root nodules.

Word Description	Morphology of Bacteroids
Rhizobium sp. in nodules	

3. Draw and describe colonies from the root nodule and also methylene blue stained preparations from the colonies and the pure culture of *Rhizobium* species.

Description of Colony	Methylene Blue Stained Smear from Colony	Methylene Blue Stained Pure Culture

DISCUSSION

1. What is the importance from an ecologic viewpoint of symbiotic nitrogen fixation?

2. Aside from legumes, what other plants fix nitrogen symbiotically? Use any reference material you desire.

PART B. AMMONIFICATION BY SOIL MICROORGANISMS

Nitrogen in bacteria, higher plants, and animals exists in the forms of nucleic acids, nitrogen-containing coenzymes, and proteins. Nitrogen in the proteins is returned to the soil in the most reduced form (NH_3) through a series of hydrolytic decompositions mediated by many different kinds of soil bacteria and fungi. The first few steps in decomposition of the native proteins are mediated by extracellular hydrolytic enzymes and collectively are often called *peptonization* (formation of water-soluble proteoses and polypeptides).* Proteoses and polypeptides are further hydrolyzed to peptides and ultimately to individual amino acids. Amino acids, in turn, are deaminated to liberate ammonia. This production of ammonia from nitrogenous organic compounds is called *ammonification*. The following generalized equations represent the ammonification process.

1. $$\text{proteins of dead microorganisms, plants and animals} \xrightarrow[H_2O]{\text{soil microorganisms}} \text{amino acids} \xrightarrow[H_2O]{\text{soil microorganisms}} NH_3 + CO_2 + H_2O \text{ or organic acids}$$

2. $$\underset{\text{urea}}{CO(NH_2)_2} \xrightarrow[H_2O]{\text{soil microorganisms}} 2NH_3 + CO_2$$

*In the past, often called peptones.

Ammonia released by ammonification may be lost into the air, utilized by other microorganisms for synthesis of their protein, or used as an energy source by certain chemoautotrophic organisms known as nitrifiers, which will be discussed in the next section. In this section we will demonstrate the rate of ammonification of protein, polypeptides (peptones), and urea by soil microorganisms.

PERIOD I

MATERIALS (for a pair of students)

1. 5 tubes of urea broth (yeast extract, 0.1 g; KH_2PO_4, 9.1 g; Na_2HPO_4, 9.5 g; urea, 20.0 g; in 1000 ml of distilled water. Filter sterilize and distribute into tubes under aseptic conditions).
2. 5 tubes of peptone broth.
3. 5 tubes of casein (milk protein) broth.
4. Nessler's reagent (see Appendix for formula).
5. 1 bottle of brom thymol blue indicator.
6. 2 samples of soil; one of good soil (G), the other of poor soil (P).
7. 1 white porcelain spot plate.

PROCEDURE

1. Obtain 5 tubes each of urea, peptone, and casein broth and label as to content.

2. Choose one soil sample to be used by you (good soil, G) and one to be used by your partner (poor soil, P). Label 1 tube of each of the 3 media as G; label 1 of each medium as P; 1 of each medium as your unknown; 1 of each as your partner's unknown; and keep the remaining set of 3 tubes as uninoculated control.

3. Inoculate the 3 tubes (urea, peptone, and casein) labeled G with 2 loopfuls of soil G. The soil can be readily transferred by wetting a flame-sterilized loop in the liquid to be inoculated. Similarly inoculate the 3 tubes labeled P with soil P. Also inoculate appropriately labeled tubes with your unknown and your partner's unknown.

4. Mix the contents of the inoculated tubes and transfer 1 drop from each tube to separate depressions on a clean spot plate for determination of 0 hr ammonia content. To each of these spots add 1 drop of Nessler's reagent. If no ammonia is present, no color will develop; if a slight amount of ammonia is present, you may observe a faint yellow color; if a large quantity of ammonia is present, a dark yellow to brown precipitate will form.

5. Record your observations immediately in the space provided. Nessler's solution is very sensitive to ammonia. (If you wait too long, a small amount of ammonia present in the air may dissolve into the sample and give false results.) Use + for a faint yellow color (small amount of ammonia), + + for deep yellow (more ammonia), + + + for brownish precipitate (large quantity of ammonia), and – for no color.

6. Incubate inoculated tubes at 30C or room temperature for 2 to 5 days.

PERIOD II

PROCEDURE

1. Repeat Steps 4 and 5 after 48 hr and 4 days. Use uninoculated media as control, i.e., test both at the same time for presence of ammonia. Do not contaminate the control tubes during the course of the experiment.

2. Clean the spot plate again; add 1 drop from each tube, including controls, to separate depressions. Add 1 drop of brom thymol blue to each spot. Compare the color produced in the inoculated tubes with the control tubes. Record the results of pH (acid or alkaline) measurements in the space provided. Use + for faint blue, + + for blue, + + + for dark blue, and – for green, which indicates neutral pH. Note, however, that in acid (pH below 6.5) brom thymol blue will turn yellow. Your instructor will provide you with pH meter readings of culture media inoculated with good and poor soil at 0 hr, 48 hr, and 4 days for comparison.

RESULTS

1. Record the rate of ammonification of urea, peptone, and casein media by soil microflora in good and poor soil and by your unknown.

Samples	Media	Ammonia Produced		
		0 hr	*48 hr*	*4 days*
Good Soil	Urea			
	Peptone			
	Casein			
Poor Soil	Urea			
	Peptone			
	Casein			
Your Unknown	Urea			
	Peptone			
	Casein			
Control	Urea			
	Peptone			
	Casein			

2. Record the changes in pH of the media measured by brom thymol blue indicator (+, + +, + + +) and pH meter readings from your instructor.

Samples	Media	0 Hr		48 Hr		4 Days	
		BTB	*pH meter*	*BTB*	*pH meter*	*BTB*	*pH meter*
Good Soil	Urea						
	Peptone						
	Casein						
Poor Soil	Urea						
	Peptone						
	Casein						
Your Unknown	Urea						
	Peptone						
	Casein						
Control	Urea						
	Peptone						
	Casein						

DISCUSSION

1. Why is the ammonification of casein slower than that of urea or peptone?

2. What is (are) the reason(s) for the change in pH of those tubes which showed high ammonia content?

PART C. NITRIFICATION BY SOIL OR AQUATIC MICROORGANISMS

Practically all of the autotrophs and heterotrophs assimilate nitrogen in combined forms. While many bacteria can assimilate this nitrogen in the form of ammonium salts, many heterotrophs, including animals, require nitrogen in amine form, as in amino acids. Higher plants, however, mainly assimilate nitrogen in its most oxidized form as nitrate. Fortunately, ammonia liberated by soil microorganisms through the ammonification process is oxidized to nitrate in soil or in water by highly specialized groups of aerobic chemoautotrophs; thus nitrogen, in the form of nitrate, is made available to the plants.

The microbial oxidation of ammonia to nitrates is called *nitrification.* The oxidation of ammonia to nitrates by chemoautotrophs (chemolithotrophs) involves two distinct steps, and each step is mediated by organisms belonging to different genera. The first step, sometimes called *nitrosification,* involves the oxidation of ammonia to nitrite and is carried out by species of the genera *Nitrosomonas, Nitrosococcus,* and others of the family Nitrobacteraceae.

$$2NH_3 + 3O_2 \xrightarrow[\textit{Nitrosococcus}]{\textit{Nitrosomonas} \text{ or}} 2HNO_2 + 2H_2O + 79 \text{ kcal}$$

Energy liberated by the oxidation is used by these organisms for their growth.

Most higher plants are unable to utilize nitrites as their source of nitrogen, and high nitrite concentrations are toxic to many plants and animals. However, members of the genera *Nitrobacter* and *Nitrococcus* and others of the family Nitrobacteraceae obtain their energy by oxidizing nitrites to nitrates.

$$HNO_2 + \tfrac{1}{2}O_2 \xrightarrow[\textit{Nitrococcus}]{\textit{Nitrobacter} \text{ or}} HNO_3 + 21.6 \text{ kcal}$$

This second step in the nitrification process thus produces an immediately available form of nitrogen for agriculture.

Nitrifiers as a rule are highly aerobic and are very sensitive to acidity. Optimum pH is around 8.0 to 8.6; consequently, activities of these organisms are reduced unless a soil is well buffered or limed. While nitrification may occur rapidly in soil under suitable conditions, it has been found that the process is relatively slow in laboratory culture media. In this experiment we will demonstrate the presence of nitrifiers in soil by utilizing media containing ammonium salt for the first step and nitrite for the second step in nitrification. Because of the lengthy period involved in the oxidation of ammonia and nitrite, your instructor will provide you with preinoculated culture media (2 to 4 weeks old) to be used in Period II. Oxidation of these compounds will be monitored and qualitatively analyzed.

PERIOD I

MATERIALS (for 2 students)

1. 2 125 ml Erlenmeyer flasks with 30 ml of ammonium medium (Medium A).

2. 2 125 ml Erlenmeyer flasks with 30 ml of nitrite medium (Medium B).

3. Samples of soil in Media A and B. (Note to instructor: 1 g of soil should be added to 100 ml of Medium A. Mix well, allow to settle, and pipette approximately 15 ml of liquid above sediment into containers suitable for distribution to students. This is to avoid scratching the Erlenmeyer flasks with the silicate in the soil during incubation on the rotary shaker. Similarly prepare inoculum for Medium B.)

4. Nessler's reagent.

5. Nitrite reagents #1 and #2 (see Appendix or *Manual of Microbiological Methods,* McGraw-Hill Book Co., Inc., 1957).

6. 1% aqueous antazoline solution for nitrate test (see Appendix).

7. 1 tube of 1% potassium nitrate solution (use as control for nitrate test).

8. 1% sulfamic acid.

9. Concentrated H_2SO_4.

10. 4 small lip test tubes.

11. 1 white porcelain spot plate.

PROCEDURE

1. Working in pairs, one partner inoculate 1 flask of Medium A with 10 ml of soil sample prepared in Medium A. Mix thoroughly. Keep the other flask as uninoculated control.

2. Do the same for Medium B, except use soil sample prepared in Medium B. Keep the second flask as uninoculated control. (Examine the composition of both media, find major differences between the 2 media.)

Ingredients	Ammonium Medium (A)	Nitrite Medium (B)
$(NH_4)_2SO_4$	2.0 g	—
$NaNO_2$	—	1.0 g
K_2HPO_4	1.0 g	1.0 g
$MgSO_4$	0.5 g	0.5 g
$FeSO_4$	0.1 g	0.1 g
$MgCO_3$	5.0 g	—
$NaHCO_3$	—	1.0 g
Distilled water	1000 ml	1000 ml

NOTE: All of the ingredients are inorganic substances. What is (are) the function(s) of each compound? Refer to Experiment 13, if necessary.

3. Test the presence or production of ammonia, nitrites, and nitrates by the following procedures.

Test for Ammonia

a. Transfer 1 drop of uninoculated (control) Medium A into a depression in a *clean* spot plate by using a sterile pipette. Use aseptic technique to avoid contamination of control.

b. Do the same in an adjacent depression for the inoculated Medium A.

c. Add 2 to 3 drops of Nessler's reagent to both samples. Formation of a brown precipitate indicates the presence of a large amount of ammonia (+ + +); deep yellow to orange color indicates a moderate amount of ammonia (+ +); a pale yellow color indicates a small amount of ammonia (+); no color indicates absence of ammonia (−). Record all of your observations in the space provided.

NOTE: It should be remembered that as Step 1 of nitrification proceeds, the residual ammonia should decrease as it is oxidized to nitrite.

Test for Nitrite

This test will be used to determine the amount of nitrite formed from the oxidation of ammonia. The test is also applicable for determining the decrease in amount of nitrite in Medium B, as a result of the oxidation of nitrite to nitrate.

a. With a sterile pipette, aseptically transfer 1 drop of inoculated Medium A into a depression of a clean spot plate.

b. Do the same for the uninoculated control and the inoculated Medium B. (Be sure to make a clear label on the edge of the spot plate for identification of the samples.)

c. Add 1 drop of nitrite reagent #1 to each of the 3 spots; wait about 30 seconds, then add 1 drop of nitrite reagent #2 to each of the 3 spots. Mix the samples and reagents by gently tilting the spot plate. Use + for faint pink color, which indicates the presence of a small amount of nitrite; + + for definite pink color, which indicates a moderate amount of nitrite; + + + for dark purplish pink, which indicates a large amount of nitrite; and − for no color, which indicates the absence of nitrite. Record all of your observations in the space provided.

Test for Nitrate

a. Aseptically transfer 0.3 ml each of uninoculated and inoculated Medium B and reagent control (1% KNO_3 solution) into 3 small test tubes provided. (Be sure to label tubes as UI = uninoculated, I = inoculated, C = control.)

b. Add approximately 0.3 ml of 1 per cent sulfamic acid; mix well, wait for approximately 30 seconds, then add 0.3 ml antazoline hydrochloride; mix well. The presence of a large amount of nitrite frequently interferes with a clear-cut nitrate test. All nitrite will be removed in this step while the nitrate present will be unaffected by the addition of sulfamic acid.

c. Hold the tubes in a slightly tilted position and add about 0.6 ml of concentrated H_2SO_4 dropwise to allow the acid to run down along the inner surfaces of the tubes. (*Do not spill or get on your hands. The acid is corrosive and harmful to the skin. Wash with water immediately if the acid is spilled or gets onto your hands.*) The object here is to make two distinct layers, the sample plus added reagent above and the sulfuric acid below. If nitrate is present, a deep red ring forms between the two layers.

NOTE: This test is primarily qualitative rather than quantitative. However, based on intensity of the color development, you may record + for faint red, + + for red, + + + for deep red, and − for no color development. Record all of your observations in the space provided.

4. Incubate inoculated flasks (Media A and B) on a shaker and keep uninoculated controls in your desk cabinet until next period.

PERIOD II

MATERIALS

1. Media A and B from last period (1 week old).

2. Samples of inoculated Media A and B, 2 to 4 weeks old. (These were preinoculated by your instructor to shorten the length of incubation for this experiment.)

PROCEDURE

1. After incubation for 1 week, 2 weeks, and 4 weeks:

   a. Medium A should be tested for reduction in ammonia content, production of nitrites, and, on 4 week cultures, for nitrates.

   b. Medium B should be tested for reduction in nitrite content and increase in nitrate concentration.

RESULTS

1. Oxidation of ammonia in Medium A to nitrite (first stage of nitrification):

Presence of	Sample	Incubation Time			
		0 hr	*1 week*	*2 weeks*	*4 weeks*
Ammonia	Control				
	Inoc.				
Nitrite	Control				
	Inoc.				
Nitrate	Inoc.	—	—	—	

2. Oxidation of nitrite in Medium B to nitrate (second stage of nitrification):

Presence of	Sample	Incubation Time			
		0 hr	*1 week*	*2 weeks*	*4 weeks*
Nitrite	Control				
	Inoc.				
Nitrate	Control				
	Inoc.				

DISCUSSION

1. What is the reason for testing the presence of nitrate in inoculated Medium A after incubating for 4 weeks? (Medium A is designed for Step 1 of nitrification.)

PART D. NITRATE REDUCTION AND DENITRIFICATION

Nitrates produced in a soil by nitrifying bacteria through the oxidation of ammonia or applied to a soil in the form of commercial fertilizer are very soluble and if not promptly assimilated by plants may be lost from the soil either by leaching or by microbial reduction.

Nitrates can be used by many different bacteria. Some bacteria are capable of reducing nitrates to ammonia and of subsequently incorporating ammonia into amino acids and proteins, the process being known as *reductive assimilation of nitrate.* Other bacteria are able to use nitrate in place of oxygen as a final hydrogen acceptor (the process being known as *reductive* or *respiratory dissimilation of nitrates*). The latter process usually occurs whenever organic matter, such as carbohydrates and proteins, is decomposed under anaerobic conditions. Under these conditions the oxidizable organic matter is called the hydrogen donor and nitrate is called the hydrogen acceptor (remember the biological definitions of oxidation and reduction).

The ability to reduce nitrate to nitrite in the absence of oxygen is possessed by many bacteria and the process is called *nitrate reduction.*

$$NO_3^- \xrightarrow[\text{anaerobic}]{2e^- + 2H^+} NO_2^- + H_2O \qquad (1)$$

However, the ability to reduce nitrate beyond nitrite is limited to a relatively small number of genera. Many species of *Pseudomonas* have this ability, reducing nitrate to gaseous nitrogen (N_2). The formation of N_2 from NO_3^- via NO_2^- is known as *denitrification.*

$$NO_3^- \xrightarrow[\text{anaerobic}]{5e^- + 6H^+} \tfrac{1}{2}N_2 + 3H_2O \qquad (2)$$

Based on schematic equations, nitrate as the electron acceptor in the nitrate reduction process (1) is of limited physiologic value; for each molecule of nitrate reduced, only two electrons can be accepted instead of five as in the denitrification process (2). Another way of expressing the same idea is that under anaerobic conditions, more food materials can be oxidized by denitrification than by nitrate reduction. In either process, microorganisms gain energy as the food material is oxidized under anaerobic conditions. Gaseous nitrogen produced by denitrification is lost to the atmosphere until it is again captured by nitrogen-fixing organisms. The ability to carry out nitrate reduction or denitrification is a unique physiologic property of microorganisms, hence it is used in characterizing bacteria.

PERIOD I

MATERIALS

1. 4 tubes of tryptone–yeast extract–nitrate broth (TYNB) with Durham insert (tryptone, 15.0 g; KNO_3, 1.0 g; yeast extract, 3.0 g; 1000 ml distilled water).
2. 1 white porcelain spot plate.
3. Nitrite reagents #1 and #2.
4. Nitrate reagents: antazoline-HCl and concentrated H_2SO_4.
5. Sample of soil.
6. Your purified unknown culture.
7. Broth culture of *Pseudomonas aeruginosa.*

PROCEDURE

1. Inoculate 1 tube of TYNB with a loopful of a soil provided and label the tube "S." Similarly inoculate 1 tube of the broth with your unknown culture, 1 tube with a pure culture of *P. aeruginosa,* and keep the fourth tube as an uninoculated control. Label all of the tubes.
2. After gently mixing the tubes by rotating them between the palms of your hands, test for nitrates and nitrites as you did in the preceding section of this experiment.
3. Incubate the tubes in your desk cabinet for 4 to 5 days.

PERIOD II

PROCEDURE

1. After 4 to 5 days of incubation, check for gas formation in the Durham inserts. Presence of gas indicates denitrification of nitrate; that is, nitrate is reduced to gaseous nitrogen via nitrite. Test for presence of nitrite as well as

nitrate. If the inoculated tubes produced no gas and no color development for either nitrite or nitrate tests, it is probable that the nitrate was reduced to nitrite and on to ammonia. However, the ammonia test with Nessler's reagent is omitted, since ammonia could arise from the reduction of nitrate or by ammonification of organic nitrogenous material (tryptone and yeast extract) in the media. Record all of your observations in the space provided.

RESULTS

1. Record your observations on nitrate reduction and denitrification. Use + and − signs to indicate presence or absence of the end products.

	0 Hr			4 to 5 Days		
Inocula	*Gas*	NO_2^-	NO_3^-	*Gas*	NO_2^-	NO_3^-
Soil						
Your Unknown						
P. aeruginosa						
Uninoculated Control						

DISCUSSION

1. Under what environmental conditions may farmers lose soil nitrate?

2. Compare nitrate reduction with denitrification.

SECTION VI

SELECTED EXPERIMENTS IN APPLIED MICROBIOLOGY

EXPERIMENT 27

Microorganisms Involved in Sewage Treatment

Sewage is the used water supplies of home, community, or industry. Although the amount of domestic sewage produced by each individual varies greatly, an average of 100 gallons (0.4 cubic meters) per person per day is considered normal for most large cities of the United States. Domestic raw sewage consists of household wastes, often augmented by garbage disposal waste, and is composed of 98.8 to 99.5 per cent water and 0.5 to 1.2 per cent solid matter.

The solid matter in sewage is classified as (1) *dissolved solid or soluble substances,* composed chiefly of mono- and disaccharides, amino acids and small peptides, fatty acids, glycerol, and many inorganic ions; (2) *colloidal and suspended solids,* made up of polysaccharides (such as starch and cellulose), proteins, fats, and insoluble inorganic compounds in colloidal state; and (3) *settleable solids,* including all materials which settle when the sewage is allowed to stand for a short period of time. The dissolved oxygen (D.O.) content of untreated sewage is usually near zero, or becomes zero when the sewage is allowed to stand for a few minutes. The pH ranges from 6.0 to 8.5. Although the amount of organic matter present in sewage appears small, sewage contains all of the nutrients necessary to nourish many kinds of bacteria.

Both quantitatively and qualitatively, the microbial flora of untreated or raw sewage vary greatly from community to community; thus the number of microorganisms present in sewage is usually expressed in ranges of between 500,000 and 20,000,000 microorganisms per milliliter. Qualitatively, these are largely intestinal bacteria, but many soil microorganisms, including algae and protozoans, are also present. It should be stressed here that most of the microorganisms of sewage are nonpathogenic and include aerobic, facultative, and anaerobic types. They are primarily heterotrophic (saprophytic) mesophiles, although some are psychrophiles, thermophiles, or autotrophic types. These organisms, including pathogens, have extremely complex metabolisms and ecological relationships in the sewage, such as *commensalism* (see Experiment 24), *synergism* (see Experiment 25), and *antagonism* (see Experiment 18).

Raw sewage discharged into a stream or a body of water may be diluted enough to appear inconsequential to the usefulness of the receiving water, but when the discharges continue and the amounts of nutrients increase, the receiving body eventually becomes overfertilized *(eutrophic).* This leads to algal bloom and other aquatic nuisances and results in the loss of the water's usefulness for fish and other aquatic life, for recreation, and as a potable water supply.

Therefore, sewage must be treated before it is discharged into a receiving body of water.

The objectives of sewage treatment are first, to remove or stabilize organic matter capable of supporting microbial growth, and secondly, to remove or destroy pathogenic microorganisms. There are numerous treatments, both biologic and chemical, which accomplish these objectives (for further discussion see your text). Properly treated sewage (effluent) contains only traces of organic materials and colloidal or suspended matter, and comparatively few bacteria.

The objectives of this experiment are to compare the total number of viable bacteria (viable count) and to test for the presence of coliform bacteria (gram-negative, nonsporeforming bacilli, which ferment lactose with acid and gas production within 24 to 48 hr when incubated at 35C) in both treated and untreated domestic sewage. Coliform bacteria are residents of the intestinal tract and are discharged by humans and other warm-blooded animals. Hence, the presence of coliforms in any given water sample is a good indication of fecal contamination and, therefore, of possible infection with enteric pathogens, such as *Salmonella typhi* (typhoid fever), *S. paratyphi* (paratyphoid fever), *Shigella dysenteriae* (dysentery), *Vibrio cholerae* (Asiatic cholera), and *Entamoeba histolytica* (amebic dysentery). The presence of coliform bacteria will be determined by the *presumptive* test, using the Most Probable Number Method (MPN), which consists of inoculating a series (usually sets of 3 or 5 tubes) of lauryl tryptose broth with a measured amount of water sample. For each sample you will use 3 sets of 5 tubes with differing dilutions. The experiment will be performed by a pair of students, each pair receiving either raw or treated sewage.

PERIOD I

PART A. TOTAL NUMBER OF VIABLE BACTERIA (VIABLE COUNT) IN SEWAGE

MATERIALS (for a pair of students)

1. A bottle of raw *or* treated sewage.
2. 6 90-ml sterile water blanks for raw sewage (RS);
   4 90-ml sterile water blanks for those pairs assigned treated sewage (TS).
3. 6 sterile Petri dishes.
4. 1 bottle of sterile, melted nutrient agar.
5. A canister of pipettes.

PROCEDURE

1. For RS, arrange the sterile 90 ml water blanks and label the bottles as 10^{-1}, 10^{-2}, 10^{-3}, 10^{-4}, 10^{-5}, and 10^{-6}. (For TS sample, dilutions end at 10^{-4}.)

2. a. Using a sterile 10 ml pipette, carefully transfer 10 ml of the sample into the bottle labeled 10^{-1}. Discard the pipette into a receptor, replace the bottle cap, and mix the contents thoroughly.

b. Pipette 10 ml from the bottle labeled 10^{-1} into the bottle labeled 10^{-2} and mix the contents of the bottle thoroughly.

c. Repeat the process until all of the labeled dilution blanks are used, 10^{-6}, for RS; 10^{-4} for TS.

3. a. For RS, pipette 1.0 ml from 10^{-5} bottle into each of 2 dishes (duplicate plate count). Label the dishes RS 10^{-5}.

b. Do the same with 10^{-6} diluted sample and label the dishes as RS 10^{-6}.

c. Carefully pipette 0.1 ml of 10^{-6} dilution into each of 2 dishes. Label the dishes as RS 10^{-7}. Note: Only 0.1 ml of sample is used; therefore the dilution is increased by 10, or 10^{-7}.

d. For treated sewage, use 10^{-3}, 10^{-4}, and 10^{-5} dilutions. Remember, 10^{-5} dilution is obtained by using 0.1 ml of 10^{-4} dilution.

4. When all dishes are inoculated, obtain a bottle of melted nutrient agar from a water bath (45 to 50C). Aseptically pour 15 to 20 ml of the agar into each dish. Mix the agar and inoculum thoroughly by gently rotating and tipping the dish. Allow the agar to solidify.

5. After the agar has solidified, invert the plates and incubate at 35C for 48 to 72 hr. 37C incubation can be used if 35C is not available, but the latter is the temperature specified in American Public Health Association, *Standard Methods for the Examination of Water and Wastewater,* 13th ed., 1971.

PART B. STANDARD TOTAL COLIFORM MPN TEST (PRESUMPTIVE TEST)

MATERIALS (for a pair of students)

1. 10 tubes of single strength and 5 tubes of double strength lauryl tryptose broth with Durham insert (lactose, 5.0 g; tryptose, 20.0 g; K_2HPO_4, 2.75 g; NaCl, 5.0 g; sodium lauryl sulfate, 0.1 g [wetting agent which gives best results as a noninhibitive selective agent for members of the coliform group; see *Difco Manual,* 9th ed., Difco Laboratories, Detroit, Michigan]; 1.0 ml of 1.6% BTB indicator, in 1000 ml distilled water). Note: Concentration of the broth and added portion of the sample must conform to the requirements given in Table 27-1.

2. Diluted raw or treated sewage from Part A.

3. A canister of pipettes.

PROCEDURE

1. Students working with RS use 10^{-6} dilution bottle. Label 5 tubes of double strength lauryl tryptose broth as 10^{-6}, 10 ml; 5 tubes of single strength

TABLE 27-1
PROPER CONCENTRATION OF INGREDIENTS FOR VARIED AMOUNTS OF INOCULUM

Amount of Inoculum Used (ml)	Amount of Medium in tube (ml)	Volume of Medium + Inoculum (ml)	Dehydrated Lauryl Tryptose Broth Required (g/l)
1 ml or less	10 or more	11 or more	35.6
10	10	20	71.2
10	20	30	53.4
100	50	150	106.8
100	35	135	137.1
100	20	120	213.6

lauryl tryptose broth as 10^{-6}, 1.0 ml; and 5 tubes of single strength lauryl tryptose broth as 10^{-6}, 0.1 ml.

2. Students working with TS follow Step 1, except use 10^{-2} dilution bottle (10^{-2}, 10 ml; 10^{-2}, 1.0 ml; 10^{-2}, 0.1 ml).

3. Using a 10 ml pipette, aseptically inoculate 10 ml of 10^{-6} dilution (RS) into a tube labeled 10^{-6}, 10 ml. Repeat the same for the remaining 4 tubes labeled 10^{-6}, 10 ml. Students using TS do the same with tubes labeled 10^{-2}, 10 ml.

4. Using a 5.0 ml pipette, aseptically remove 5.0 ml of 10^{-6} dilution (RS) and dispense 1.0 ml into each of the 5 lactose broth tubes labeled 10^{-6}, 1.0 ml. Students with TS do the same, except your tubes are labeled 10^{-2}, 1.0 ml.

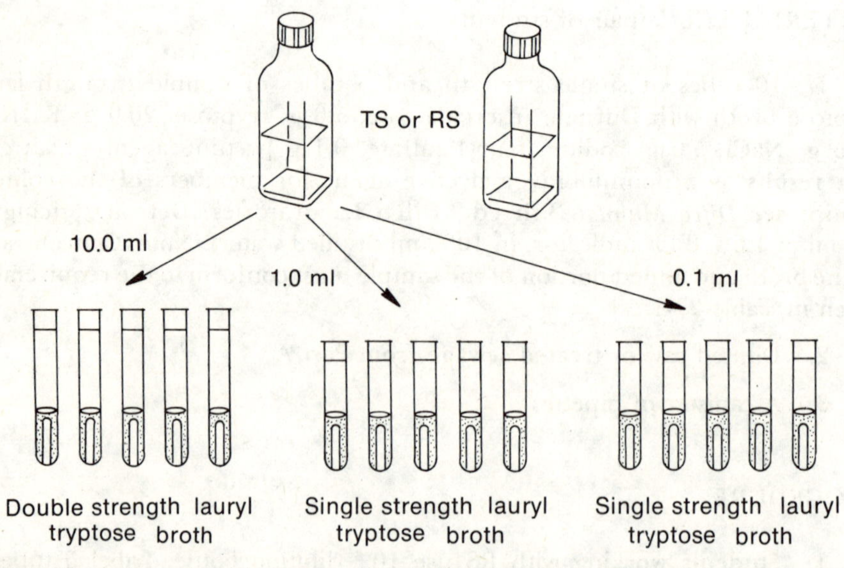

FIGURE 27–1. Most Probable Number Test (MPN). Inoculation of lauryl tryptose broth with diluted raw sewage or treated sewage.

5. Using 1.0 ml pipette, inoculate the remaining 5 tubes of lauryl tryptose broth with 0.1 ml of the diluted sample (see Fig. 27-1).

6. Incubate the inoculated tubes at 35C and examine for gas and acid production after 24 hr and again at the end of 48 hr.

PERIOD II

PROCEDURE

PART A. VIABLE COUNT

1. After incubation for 48 to 72 hr, examine and count colonies on all plates. If colony count exceeds 300, record the count as TNC (too numerous to count). Record all counts in the space provided.

2. Calculate the number of viable organisms per ml of the original sample and record in the space provided. For this calculation, use only the pair of plates (inoculated with the same dilution) having counts between 30 and 300 colonies each. Why is this important?

PART B. PRESUMPTIVE TEST: MPN DETERMINATION

1. After incubation for 24 ± 2 hr and 48 ± 3 hr, observe the change in brom thymol blue indicator (BTB). Yellow indicates acid production (A); blue indicates alkaline (Alk); record no change in color as NC. Record your observations in the space provided.

2. Examine the Durham tubes of lauryl tryptose broth for gas production. A positive presumptive test is one indicating gas production, usually within 24 hr. Broth may be saturated with gas. Give a gentle tap to the bottom of each tube which shows acid but no gas production. If the broth is saturated, tapping may cause rapid evolution of gas which may quickly become trapped in the Durham tube. If evolution of gas occurs after 24 hr, the test is generally considered doubtful presumptive; if no gas is produced after 48 hr, the test is considered negative presumptive.

3. The most probable number (MPN) of coliforms present in a sample can be estimated by use of the MPN Index (see Appendix). Various combinations of positive and negative results from five 10-ml portions, five 1.0-ml portions, and five 0.1-ml portions are used. Your instructor will explain the use and statistical significance of the number.

4. Compute MPN of coliforms/100 ml of the original sample by using the MPN table and record the results in the space provided.

5. Keep a lauryl tryptose broth tube with gas formation from the highest dilution for confirmed and completed tests in Experiment 28.

RESULTS

PART A. VIABLE COUNT

1. Total number of viable bacteria (viable count).

a. Raw sewage:

	10^{-5} Plates			10^{-6} Plates			10^{-7} Plates		
Sample	1	2	*Average*	1	2	*Average*	1	2	*Average*
RS									

The number of viable organisms per ml of raw sewage is ________cells/ml.

b. Treated sewage:

	10^{-3} Plates			10^{-4} Plates			10^{-5} Plates		
Sample	1	2	*Average*	1	2	*Average*	1	2	*Average*
TS									

The number of viable organisms per ml of treated sewage is ________cells/ml.

PART B. PRESUMPTIVE TEST: MPN DETERMINATION

1. Change of BTB indicator to yellow color and gas production in Durham tubes. (Express as 5/5 or 0/5, etc., on lower line.)

		10.0 ml Inoculum	1.0 ml Inoculum	0.1 ml Inoculum
RS	Acid			
	Gas			
TS	Acid			
	Gas			

2. Most probable number of coliforms/100 ml in the original raw sewage is ________________.

Most probable number of coliforms/100 ml in the original treated sewage is ________________.

DISCUSSION

1. Define presumptive test.

2. Why do we test for the presence of coliforms rather than for pathogenic microorganisms?

3. List the advantages and disadvantages of coliforms as an indicator of fecal pollution.

EXPERIMENT 28

Bacteriologic Examination of Water

Natural waters usually contain not only large number of microorganisms but also a wide variety, and such waters may be perfectly safe to drink. The sanitary quality of drinking (potable) water is determined primarily by the kinds of microorganisms present rather than by the count. Hence, water from streams, springs, wells, rivers, ponds, and lakes which contains a multitude of microorganisms is considered potable as long as pathogens for humans and injurious chemicals are absent.

Public Health Service Drinking Water Standards (1973) states in summary that drinking water should be entirely free from pathogenic microorganisms. However, routine examination of water for the presence of all human pathogens would be a time-consuming, tedious, and difficult—if not impossible—task. Consider the problem of trying to identify obligate intracellular organisms such as pathogenic viruses and rickettsias. The intestinal pathogens frequently transmitted by contaminated water are the causative agents of typhoid and paratyphoid fevers, bacillary dysentery, urogenital infectious diseases, and, in certain areas of the world, cholera *(Vibrio cholerae).* They are disseminated by the feces of patients or healthy carriers. Human fecal material is carried away by water in sewerage systems that may discharge into lakes and rivers supplying drinking water for other communities. Since the pathogens are of fecal origin and are introduced into potable water via sewage pollution, they can be guarded against if sewage is detected by a bacteriologic test designed to locate organisms indigenous to the intestinal tract of man and rarely found in other environments. *Coliform bacteria* are used as the indicator organisms because sewage always contains them in large numbers. Furthermore, they can be detected more easily and quickly than most pathogens, enabling public health officials to act speedily to stop the consumption of potentially contaminated water. *Streptococcus faecalis* is also often used an indicator organism in conjunction with coliforms. It should be emphasized, however, that since most coliforms are nonpathogenic, their detection in drinking water merely indicates the presence of fecal matter, not necessarily that of harmful microorganisms.

Currently, coliform organisms are further divided into two types on the basis of incubation temperature. *Total coliforms* are those organisms which ferment lactose with the production of acid and gas within 48 hours at 35C, or which form colonies with golden green sheen on membrane filters using m-Endo broth when incubated at 35 ± 0.5C for 22 to 24 hours. *Fecal coliforms,* on the other hand, are those coliform organisms able to grow and ferment lactose with production of acid and gas, or to produce blue colonies on membrane filters using m-FC broth, when incubated at 44.5 ± 0.2C for 24 ± 2 hr. Because it has been found that total coliform organisms can be isolated from sources other than fecal material, the United States Public Health Service considers fecal coliform counts more indicative of fecal pollution by man and other warm-blooded animals than counts of total coliforms. However, the counting procedure is recommended only for raw water and not for water treated for drinking purposes.

Traditionally, tests for analysis of potable water are conducted in sequence: (1) *presumptive,* (2) *confirmed,* and (3) *completed* tests. The presumptive test, as performed in Experiment 27, consists of inoculating a known amount of water sample into multiple lauryl tryptose broth tubes. If the lauryl tryptose broth is fermented with production of acid and gas within 24 to 48 hr when incubated at 35C, the water is assumed or presumed to be contaminated with fecal material and to be unsafe to drink.

Occasionally a positive result from the presumptive test is not necessarily caused by coliforms, but rather by species of the genera *Clostridium* or *Bacillus.* For this reason the *confirmatory* test is performed. The test is a simple one: all tubes of the two highest dilutions of the original sample showing gas formation within 24 to 48 hours are selected, and a loopful from each tube is inoculated into a tube of brilliant green lactose bile broth, a medium more selective for coliform detection (*Difco Manual,* 9th ed., p. 37). The inoculated bile broth tube is incubated at 35 ± 0.5C for 48 ± 3 hr. After the incubation period, if the tube shows production of gas, the confirmed test is considered positive. An alternative procedure is to make a streak culture from the positive and doubtful (small amount of gas in the Durham insert) tubes of lauryl tryptose broth onto eosin methylene blue (EMB) agar or Endo agar; the plates are incubated at 35C for 24 hr. If typical coliform colonies—colonies with greenish metallic sheen *(E. coli),* or pink colonies with dark center, sometimes described as "fish eye" colonies *(E. aerogenes)*—are formed on the EMB or Endo agar, the test indicates that gas in the presumptive tubes was most likely produced by coliforms.

If the confirmed test is positive, the completed test is necessary. One usually selects a well-isolated, typical coliform colony (described above) and inoculates a portion of the colony onto a nutrient agar slant and a portion of the colony into lauryl tryptose broth. After 24 hr incubation at 35C, a gram-stained slide is made from the slant. If the organisms selected are gram-negative, nonsporeforming bacilli, and if they produced acid and gas from lactose, the completed test is positive; i.e., the organisms present in the original sample are considered to be coliforms.

These methods are presented in detail in *Standard Methods for the Examination of Water and Wastewater,* 13th ed., American Public Health Association, 1971, pp. 635-711. (See demonstration table for a copy.)

While the above tests (presumptive, confirmed, and completed) are usually applied to drinking water supplies, *membrane filter coliform count* (MFCC, which indicates total coliform count made at 35C) and *membrane filter fecal coliform count* (MFFCC, made at 44.5C) methods are now approved by the United States Public Health Service as reliable tests for the detection of total coliforms and fecal coliforms in water in general.

In the membrane filter technique, the bacteria are retained on the filter and cultured (Fig. 28-1). The membrane filters have pores averaging 0.45 μm in

MEMBRANE FILTER METHOD

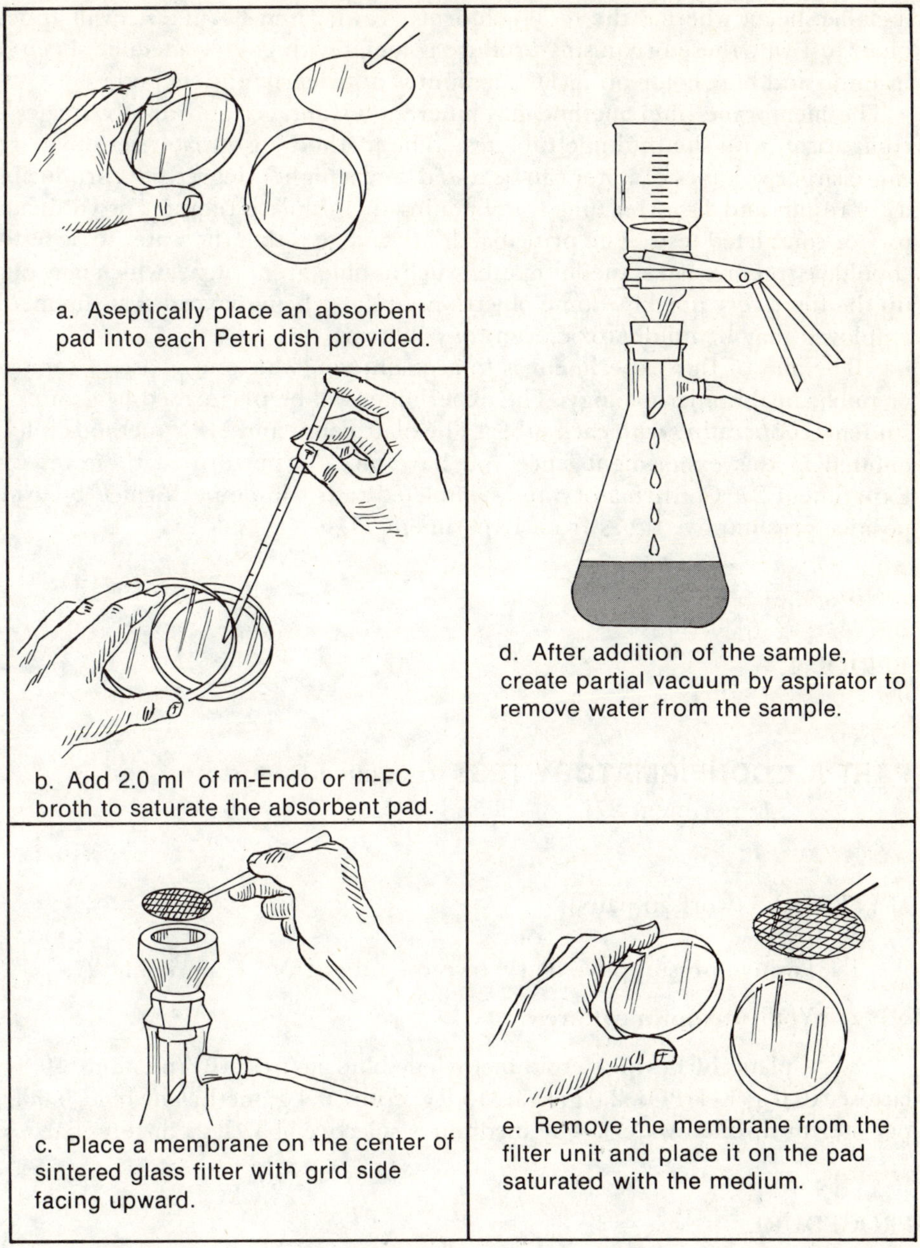

FIGURE 28–1

diameter; hence bacteria larger than 0.47 μm* cannot pass through the pores. The filter with the retained bacteria is placed on an absorbent pad saturated with liquid media, such as m-endo or m-FC broth. The coliform bacteria on the filter will form colonies that can be counted with the unaided eye or with a dissecting

*Even smaller bacteria are usually retained because of the tortuous pathway through the filter. For further discussion on membrane filters, refer to your text.

microscope. The total coliforms grown on m-Endo (35C) will exhibit golden metallic sheen, whereas the fecal coliforms grown on m-FC (44.5C) will appear blue. Just why these organisms produce colonies with golden metallic sheen on m-Endo and blue color on m-FC medium is not clearly understood.

The membrane filter method has inherent advantages and disadvantages in comparison with the multiple tube test. The method has greater sensitivity because larger volumes of water can be tested, and a higher degree of reproducibility of results and a shorter time for obtaining them since it requires no confirmatory or completed tests. The principal disadvantage is that the water to be tested should be free from extraneous matter, such as blue-green algae, which may plug up the filter very quickly. Some objections are also being voiced that the media employed may be mildly toxic even for coliforms.

The aim of this experiment is to acquaint you with one phase of sanitary or public health microbiology. The experiment will be performed by groups of students cooperating with each other. The plate count and MPN methods will be omitted in this experiment since you have already performed these tests in Experiment 27. Confirmatory and completed tests will be performed by using positive presumptive tubes from Experiment 27.

PERIOD I

PART A. CONFIRMATORY TEST (continued from Experiment 27)

MATERIALS (work in pairs)

1. Positive presumptive lauryl tryptose tubes from Experiment 27.
2. Your unknown culture.
3. 4 plates of Levine's eosin methylene blue agar (EMB) (peptone, 10.0 g; lactose, 10.0 g; K_2HPO_4, 2.0 g; agar, 15.0 g; eosin, 0.4 g; methylene blue, 0.065 g per liter of distilled water). The medium is selective as well as differential.

PROCEDURE

1. Obtain lauryl tryptose broth tubes with acid and gas formation from Experiment 27. If your sample failed to produce acid and gas from the broth, borrow the positive tubes from your neighbor. (Remember to select the highest dilution which produced acid and gas.)
2. With a sterile loop, remove one loopful of the growth from the lauryl tryptose broth tube and make a streak on the EMB agar (see Fig. 10-3 for the streak pattern). Do not cut into the agar medium!
3. Do the same with your unknown culture.
4. Invert and incubate streaked EMB agar plates at 35C for 48 hr.

PART B. MEMBRANE FILTER METHOD

MATERIALS (work in groups of 4 students)

1. Samples of surface water from above and below the treated domestic sewage outfall.

2. 3 sterile dilution water blanks (90 ml/bottle). (4 bottles for those groups assigned samples from below outfall.)

3. Membrane filter apparatus:
    a. 1 liter suction flask.
    b. A package of membrane filters and pads (10/pkg).
    c. Membrane filter clamp, forceps, funnel, and filter adapter.
    d. 10 ml m-Endo broth (for total coliform).
    e. 10 ml m-FC broth (for fecal coliform).
    f. A small beaker of 95% alcohol.
4. 8 sterile small Petri dishes.
5. 2 plastic bags.
6. 3 sterile water blanks (rinse water).
7. A canister of pipettes.

PROCEDURE

1. Prepare serial dilutions of the assigned sample as follows:
    a. Aseptically pipette 10.0 ml of the sample into a 90 ml sterile water blank and mix thoroughly. Label the bottle 10^{-1}.
    b. With a new pipette, transfer 10.0 ml sample from the 10^{-1} dilution bottle and put into a second 90 ml water blank. Mix thoroughly and label the bottle 10^{-2}.
    c. Similarly prepare 10^{-3} dilution. Those groups assigned a water sample from below the sewage outfall should make a 10^{-4} dilution also.
2. a. For those assigned samples from above the sewage outfall: label all 8 Petri dishes with A ("above"). Four of the dishes are to be labeled MFCC and will receive m-Endo broth; 4 are to be labeled MFFCC and will receive m-FC broth. Arrange the sets of 4 dishes in a row and label the first dish of each set as FS, the second dish of each set as 10^{-1}, the third dish of each set as 10^{-2}, the fourth dish of each set as 10^{-3}.

b. For those assigned samples from below the sewage outfall: label all 8 dishes with B ("below"); label with MFCC and MFFCC as above. Omit FS, label the first dish of each set as 10^{-1}, then continue with 10^{-2}, 10^{-3}, and 10^{-4}.

3. With an alcohol-flamed forceps, transfer a sterile absorbent pad to each of the 8 labeled Petri dishes.

4. Aseptically assemble a membrane filter unit as shown on the demonstration table (see Fig. 28-1).

5. Add 2.0 ml of m-Endo broth to each of the 4 dishes labeled MFCC. Add 2.0 ml of m-FC broth to each of the 4 dishes labeled MFFCC.

6. With alcohol-flamed forceps, place a membrane filter (grid side up) on the filter holder base.

7. Place the funnel on top of the membrane filter and secure it to the base with the clamp.

8. a. Attach the rubber hose to the water aspirator (vacuum source).

b. Groups assigned sample "A" use 10^{-3} dilution; groups assigned sample "B" use 10^{-4} dilution. Pipette 20.0 ml of proper dilution into funnel (use 10 ml pipette twice). (Note: the amount of water used will depend on water quality, such as turbidity and amount of particulate matter present, but ideally no less than 50 ml should be used.)

c. Start the vacuum by running the tap water.

d. Rinse the inner side of the funnel with 20 ml of sterile water (use 10 ml pipette) to remove any bacteria which may adhere to the side of the funnel and continue to evacuate until all liquid has been filtered.

e. Disconnect the vacuum source *without turning off the tap water.* Remove the funnel and carefully transfer the filter (grid side up) with alcohol-flamed forceps to Petri dish labeled MFCC 10^{-3} for Sample A and to dish labeled MFCC 10^{-4} for Sample B.

f. Repeat the same procedure with the same dilution and transfer the filter to the dish labeled MFFCC 10^{-3} (or 10^{-4}). Do not allow the base of the funnel to touch any object throughout the remaining steps of the experiment.

g. Before proceeding to the next dilution, reconnect the vacuum line and rinse the filter funnel with sterile distilled water.

9. Disconnect the vacuum source and repeat the same operation with the remaining dilutions. What is the reason for starting from the highest dilution?

10. After completion of filtration, place the 4 Petri dishes labeled MFCC in a plastic bag provided, seal to prevent dehydration of the broth, and incubate at 35C for 24 hr. Do the same with dishes labeled MFFCC, incubate at 44.5C for 22 to 24 hr.

PERIOD II

PART A. CONFIRMATORY TEST

MATERIALS

1. 2 nutrient agar slants.
2. 2 tubes of lauryl tryptose broth.

PROCEDURE

1. Observe streak culture on EMB agar. Record your observations in the space provided. Refer to page 252 for a description of coliform colonies, if necessary.

2. Did your unknown culture grow on the medium? If so, record appearance of colonies in the space provided.

3. Select one colony believed to be *E. coli* and one colony of *E. aerogenes* from your streak plate.

4. With a sterile wire needle, transfer a small amount of growth (from the *E. coli* colony) onto a nutrient agar slant. Using the same colony, transfer a small amount of growth to a tube of lauryl tryptose broth. Do the same with an *E. aerogenes* colony. Be sure that your tubes are properly labeled.

5. Incubate both the lauryl tryptose broth tubes and nutrient agar slants for 24 hr at 35C.

PART B. MEMBRANE FILTER METHOD

PROCEDURE

1. Examine the colonies on the membrane filter.
   a. Dishes labeled MFCC: count the colonies on the filter with a dissecting microscope using reflected light. Ignore all colonies that lack metallic sheen. Record your results in the space provided.
   b. Dishes labeled MFFCC: count only blue colonies (ignore all other colored colonies, such as pink and green). Record your results in the space provided.

2. Compute the number of coliforms using dishes having colony number between 20 and 80 for MFCC and between 20 and 60 for MFFCC. Since colony counts are expressed as number per 100 ml, you must multiply your count by 5 because only 20 ml of sample were used. Record your results in the space provided.

PERIOD III

PART A. COMPLETED TEST

1. Examine the lactose broth for acid and gas formation and record your observations in the space provided.

2. Prepare Gram stained slides from the nutrient agar slants. Record your observations in the space provided. If the specimen is a gram-negative nonsporeforming rod and acid and gas were produced from the lauryl tryptose broth, you have a positive completed test.

RESULTS

NOTE: Figure 28–2 summarizes the presumptive, confirmed, and completed tests.

PART A. PRESUMPTIVE, CONFIRMED, AND COMPLETED TESTS

1. Give a word description of the colonies on EMB agar.

   a. Streak made from positive presumptive test.

   b. Streak made from your unknown (for lactose fermentation, refer to your notebook, p. 135).

SUMMARY OF PRESUMPTIVE, CONFIRMED, AND COMPLETED TESTS

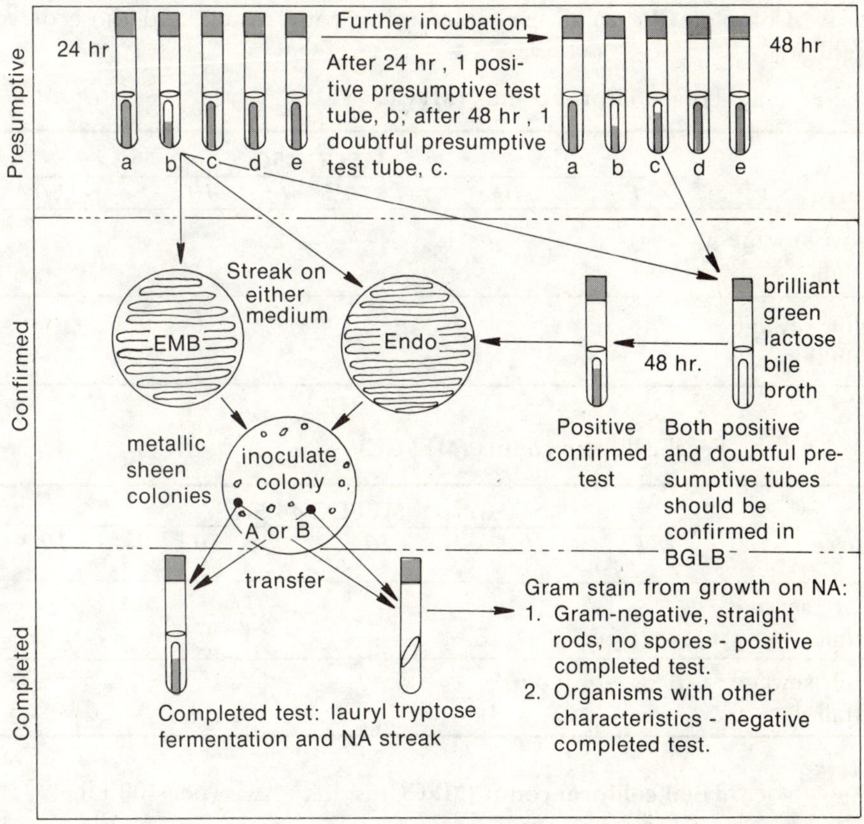

FIGURE 28–2. IMViC Test may be performed after completed test to differentiate between typical *E. coli* and *E. aerogenes.* For complete description of IMViC test, please refer to your text. IMViC is a mnemonic device; I stands for indole reaction, M for methyl red reaction, V for acetyl-methyl-carbinol test (originated by Voges-Proskauer), i is for the sake of euphony, C stands for growth in mineral solution containing citrate as a sole source of carbon.

2. Fill in the following table for confirmed and completed tests.

Culture	Stained Slides			Lauryl Tryptose Broth	
	Cell Shape	*Spores Present?*	*Gram Reaction*	*Acid*	*Gas*
Greenish, Metallic Sheen Colony *(E. coli)*					
"Fish eye" Colony *(E. aerogenes)*					

PART B. MEMBRANE FILTER METHOD

1. Complete the following tables using your results and those of your neighbor.

a. Total coliform counts (MFCC):

Source	MFCC 35C				
	FS	10^{-1}	10^{-2}	10^{-3}	10^{-4}
Above sewage outfall					
Below sewage outfall					

b. Fecal coliform counts (MFFCC):

Source	MFFCC 44.5C				
	FS	10^{-1}	10^{-2}	10^{-3}	10^{-4}
Above sewage outfall					
Below sewage outfall					

c. Total coliform count (MFCC) is ________ per 100 ml.

d. Fecal coliform count (MFFCC) is ________ per 100 ml.

DISCUSSION

1. What pathogenic organisms are most likely to be transmitted via water?

2. What is the function of the completed test? Is this test necessary?

3. List characteristics which are desirable for sewage pollution indicator organisms.

4. Explain the difference between the MFCC and MFFCC tests.

EXPERIMENT 29

Microbiologic Examination of Foods

Food microbiologists are concerned with the kinds and numbers of microorganisms present in or on food. They are concerned with prevention of disease transmission by food, with microorganisms responsible for spoilage, and also in the exploitation of useful microorganisms for the production of various food products.

Spoilage of foods is largely dependent on their physical and chemical composition (pH, moisture content, and percentage of carbohydrates, proteins, and lipids) as well as on the numbers and kinds of microorganisms (such as saccharolytic, proteolytic, and lipolytic organisms) present. Another factor involved in food spoilage is the manner in which foods are processed (raw or cooked, addition of microbial inhibitors, drying, increased salt or sugar content, etc.) and stored (canned or bottled, refrigerated or frozen, etc.). These factors will determine the kinds of organisms—aerobic or anaerobic, psychrophilic or thermophilic—which can grow and which can perhaps cause deterioration of food quality or even illness or death of man when ingested. Food poisoning, for instance, may be induced by certain species and strains of *Staphylococcus* (food intoxication), *Salmonella* (infectious gastroenteritis and fever), and *Clostridium* ("botulism," a food intoxication which is frequently lethal).

Spoilage of perishable foods such as fresh meat, fruits, and vegetables may be readily recognized by changes in the appearance, texture, color, smell, and taste of the foods. Likewise, spoilage of certain heat-processed foods, such as canned vegetables, meats, and fish, may often be detected by either a change in the acidity of the food (flat sour), or the production of H_2S (sulfide or "stinker"). Among the most easily recognized signs of canned food spoilage is change in the shape of the can resulting from gas production. As gas production begins, the ends of the can may bulge slightly or one end will bulge out if the side of the can is tapped. The ends of the can may be easily pressed in by finger pressure; hence the can is called a "flipper." If one end of the can bulges noticeably outward, it is called a "springer," but if it can be pushed in by force without bulging out at the other end, it is called a "soft swell." When gas pressure is so great that no part of the can may be pushed inward, it is called a "hard swell"; if the can remains flat and a sour flavor develops, it is referred to as "flat sour." These four types of canned food spoilage are usually caused by anaerobic, sporeforming, obligate thermophiles primarily belonging to the genus *Clostridium;* consequently, the spoilage is often referred to as Thermophilic Anaerobe, or T.A., spoilage.

Among the nonthermophilic clostridia (grow well from 20 to 35C), *C. botulinum* is perhaps of greater interest since it is the causative agent of botulism. The toxins produced by the organism are among the most potent poisons known, but the toxins are heat labile, that is, they are destroyed by boiling for 10 minutes. There are six toxigenic types (A to F) of *C. botulinum.* As a rule, types A, B, E, and F cause botulism in man, and types C and D cause the disease in animals. The endospores of the organisms are heat resistant, as in other spore-formers, but the organisms grow well on ordinary media only under strict anaerobic conditions. All types liquefy gelatin, ferment glucose, and frequently produce a foul odor with gas bubbles from cooked-meat medium. A detailed discussion of botulism is presented in your text.

As a rule, microorganisms found in or on food originate from air, utensils, and hands of man. The interior of uninjured, disease-free fruit, vegetables, or deep tissues of animals (meat) should be sterile or virtually free of microorganisms. Cleanliness and asepsis are obviously of great importance in the processing of foods. The purpose of this experiment is to give the student an indication of the numbers and kinds of organisms which may be present in certain foods.

PERIOD I

PART A. ENUMERATION OF BACTERIA IN GROUND BEEF AND HALOPHILIC ORGANISMS IN SALT-CURED MEAT

MATERIALS (work in pairs)

1. Samples of ground beef and ground salt-cured meat in Petri dishes.
2. Your unknown, your partner's unknown.
3. 2 sterile water blanks (99.0 ml/bottle).
4. 3 sterile water blanks (9.0 ml/tube).
5. 2 bottles of melted nutrient agar.
6. 3 tubes each of NB (nutrient broth), NB + 3%, NB + 6.5%, and NB+ 10% NaCl.
7. 6 sterile Petri dishes.

PROCEDURE

1. You and your partner will work with either ground beef or ground salt-cured meat, whichever is assigned. Weigh 1.0 g of the ground beef or salt-cured meat and put into 99.0 ml sterile water blank. Mix thoroughly and vigorously. Allow the meat to settle. Label the bottle 10^{-2}.

2. Place tip of pipette above the settled ground meat to avoid plugging of the tip. Aseptically pipette 1.0 ml of the 10^{-2} dilution of ground beef or salt-cured meat into a 9.0 ml sterile water blank. Mix thoroughly and label the tube 10^{-3}. Continue to dilute the suspension serially to 10^{-5} using the remaining two 9.0 ml dilution tubes.

3. Place 1.0 ml of 10^{-3} dilution into each of 2 sterile Petri dishes; do the same for 10^{-4} and 10^{-5} dilutions and label the dishes accordingly.

4. Obtain melted, cooled nutrient agar (45C) from the water bath, and dispense 15 to 20 ml into each Petri dish. Mix the agar and inoculum by carefully rotating and tilting the dish back and forth. Allow to solidify. Invert the plates and incubate for 48 hr at 37C.

5. Arrange NB and NB + NaCl media (to promote the growth of osmophilic or halophilic organisms) in a tube rack. Inoculate a set of tubes with your unknown, a set with your partner's unknown, and the other set with a small amount of salt-cured meat (obtain from a neighboring group if not assigned salt-cured meat). Use a sterile wire loop for handling the meat. Incubate the tubes for 2 to 5 days in your desk cabinet.

PART B. EFFECTS OF LOW PH AND HIGH SUGAR CONTENT ON MICROBIAL GROWTH

MATERIALS (work in pairs)

1. 5 tubes of NB (control).
2. 5 tubes of NB + citric acid (pH 3.0).
3. 5 tubes of NB + 20% glucose.
4. 5 tubes of NB + 20% glucose + citric acid (pH 3.0).
5. 1 24 hr broth culture each of *Pseudomonas fluorescens, Saccharomyces cerevisiae, Aspergillus niger,* your unknown, and your partner's unknown.

PROCEDURE

1. Arrange NB (control), NB + citric acid (pH 3.0 selective for acidophilic organisms), NB + 20% glucose (selective for saccharolytic or osmophilic organisms), and NB + 20% glucose + citric acid (pH 3.0) tubes in order.

2. Label 1 set (one tube of each medium) as P.f., the second set as S.c., the third set as A.n., the fourth set as your unknown, and the fifth set as your partner's unknown. Inoculate the set labeled P.f. with 1 loopful of *P. fluorescens.* Likewise inoculate the second set with *S. cerevisiae,* the third set with *A. niger,* the fourth set with your unknown, and the fifth set with your partner's unknown.

3. Mix the contents of the tubes and incubate in your desk cabinet for 2 to 5 days.

PART C. OBSERVATION OF THERMOPHILIC ANAEROBE SPOILAGE IN CANNED GOODS

MATERIALS (work in groups of 4 or more students)

1. Soil and weighing paper (bibulous).
2. Ice pick.
3. Cheesecloth.
4. Soldering gun and solder (low melting, 50/50).
5. Either canned creamed corn or beans.
6. Canister of pipettes.
7. Sterile cotton, 70% alcohol.

PROCEDURE

1. Clean the end of the can with sterile cotton and 70 per cent alcohol.

2. Puncture the can of creamed corn or beans (whichever assigned) with an alcohol-flamed ice pick. Carefully inoculate the can with 0.5 g of soil via the punctured hole.

3. Seal the puncture hole with an electric soldering gun. (Caution: avoid burning your hands as well as the table top. When you finish soldering, cool the gun on an asbestos pad.) Wrap the can in cheesecloth and incubate for 2 to 5 days at 55C (to promote the growth of thermophilic organisms). Observe the change in appearance of the can daily for the next 2 to 5 days (assign a student for each day).

PERIOD II

PROCEDURE

PART A. ENUMERATION OF BACTERIA IN GROUND BEEF AND HALOPHILIC ORGANISMS IN SALT-CURED MEAT

1. Examine the pour plates prepared from serially diluted ground beef or salt-cured meat suspension. Describe the kinds of colonies which appeared on the pour plates.

2. Compute the number of viable organisms per gram of the products. Use the average from each pair of plates, but count only plates having between 30 and 300 colonies. Record your results in the space provided.

3. Observe the growth of salt tolerant (halophilic) organisms. Compare the turbidity of the broth with the control. Most bacteria will not tolerate salt content above 6.5 per cent. For this reason, the salt tolerance test is frequently used as a criterion for taxonomic studies. Record your observations in the space provided.

PART B. EFFECTS OF LOW PH AND HIGH SUGAR CONTENT ON MICROBIAL GROWTH

1. Using the amount of growth in the untreated NB tube as control, compare the growth of each organism in low pH, high sugar, and low pH + high sugar broth. Record your observations in the space provided. Express the amount of growth in terms of +, + +, + + +, or −. As a rule, most bacteria grow best near pH 7, molds at pH 2 to 3, and yeasts between pH 3 and 5.

2. If growth did occur, observe and record the position (top, bottom, or throughout the broth) of growth in respect to aerobic (A), anaerobic (An), or facultative (F). Record your observations in the space provided.

PART C. OBSERVATION OF THERMOPHILIC ANAEROBE SPOILAGE IN CANNED FOODS

1. Describe the change in shape of the can in the space provided.

2. Note any odor associated with spoilage. Prepare a spore-stained slide (p. 35) from "hard swell" of T.A. spoilage. Sketch the microscopic appearance of the slide in the space provided.

RESULTS

PART A. ENUMERATION OF BACTERIA

1. a. Describe the kinds of colonies present on the pour plates prepared from the ground beef or salt-cured meat suspension.

   b. Number of bacteria per gram of ground beef or salt-cured meat is

   ______________cells/g.

2. Record growth characteristics observed from tubes of nutrient broth supplemented with NaCl.

	Your Unknown	Salt-Cured Meat
NB (Control)		
NB + 3% NaCl		
NB + 6.5% NaCl		
NB + 10% NaCl		

PART B. EFFECTS OF LOW PH AND HIGH SUGAR CONTENT ON MICROBIAL GROWTH

1. Fill in the following table (use +, + +, + + +, or − for amount of growth).

	P. fluorescens	*S. cerevisiae*	*A. niger*	Your Unknown
NB (Control)				
NB + Cit. A (pH 3.0)				
NB + 20% glucose				
NB + Cit. A + 20% glucose (pH 3.0)				

2. Is your unknown acidophilic? Saccharophilic? Explain briefly.

3. a. Growth of the following organisms occurred at (surface, bottom, or throughout):

   P. fluorescens ______________________________

   S. cerevisiae ______________________________

   A. niger ______________________________

   Your unknown ______________________________

   b. Explain why these organisms grew at certain locations in the tube.

PART C. OBSERVATION OF THERMOPHILIC ANAEROBE SPOILAGE IN CANNED FOODS

1. Describe the T.A. spoilage of either canned creamed corn or beans.

2. Sketch a representative microscopic field of spore-stained smears.

Location of endospore:

_____Central

_____Subterminal

_____Terminal

_____Irregular

DISCUSSION

1. Spoilage of canned foods is caused primarily by bacteria and not by molds. Why?

2. If lemonade (pH 3) is kept in the open air, what types of microorganisms may grow? Where? Why?

3. When ground beef suspension is subjected to membrane filter method using m-FC broth and incubated at 44.5C with appearance of blue colonies on the filter, what public health significance does this finding have?

EXPERIMENT 30

Bacteriologic Examination of Milk

Nutritionally, milk is considered one of the most complete natural foods for man and animals and thus is consumed in large amounts. Milk presents two major problems for public health officials and microbiologists. First, milk may, if not properly pasteurized, transmit several important animal diseases to man, for example, tuberculosis, brucellosis, and Q fever. Secondly, milk serves as an excellent culture medium for many types of microorganisms (bacteria, yeasts, molds) since it has a neutral pH (6.7) and contains proteins, carbohydrates, and lipids.

Average Composition of Cow's Milk	
Lactose	5.0%
Fat	3.6
Casein	2.5
Albumin and globulin	0.7
Ash	0.7
Water	87.5

Fecal contaminants or other microorganisms, including pathogens, introduced into milk via dirty hands of dairymen, dirty utensils, flies, unsanitary conditions in milkhouse, or improper transport may rapidly lower the quality of milk and create a health hazard. The contamination of milk with harmful microorganisms by the milk processors *after* pasteurization is of particular concern. Can you reason why?

Even milk collected and handled under the very best of conditions, however, will contain low numbers of a variety of microorganisms, often referred to as the "indigenous" flora. These microorganisms get into the milk in one of three ways: as airborne contaminants, as residual microorganisms in or on milk handling equipment and not killed during the washing and disinfection cycle, or as part of the resident flora present at all times in the teat canals of cows. Since pasteurization is not a sterilization procedure, many of the indigenous organisms survive this heat treatment and will multiply if given the opportunity.

When either raw or pasteurized milk is allowed to incubate at temperatures of 25 to 30C, we generally see a succession of different microbial populations appearing as first one substrate is consumed, then another, and the growth conditions are varied (see Fig. 30-1). Milk sugar (lactose) will first be fermented by many indigenous microorganisms of raw milk (especially species of genera *Streptococcus* and *Lactobacillus,* which produce varieties of acids) until the con-

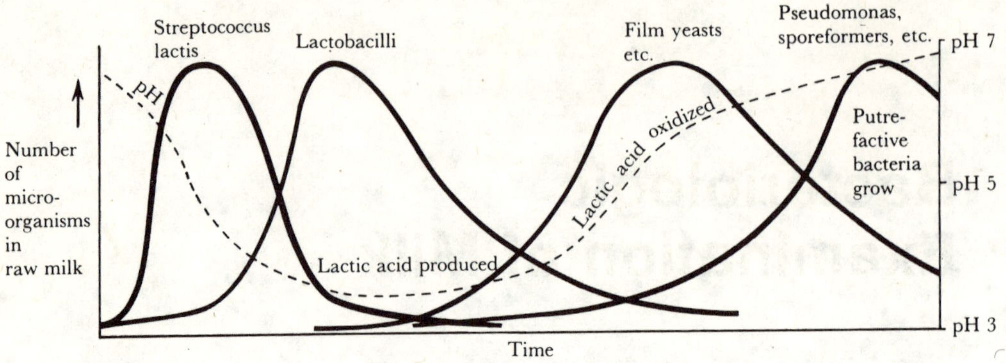

FIGURE 30–1. Sequence of microbial activities in raw milk at a moderate temperature. *S lactis* and lactobacilli multiply rapidly, ferment lactose, and produce sufficient lactic acid to bring the reaction to pH 3.5 or lower. Film yeasts, molds, etc., then oxidize the acid, and the pH rises, which permits putrefactive bacteria to grow. (From Carpenter, P. I. *Microbiology*, 3rd ed. Philadelphia: W. B. Saunders Co., 1972, p. 439.)

centration of titratable acid is such (2 to 4 per cent lactic acid) that it prevents the growth of other bacteria. Acidified milk promotes the growth of acidophilic microorganisms, including yeasts and molds. These organisms (often aerobic) in turn metabolize the acid and return the milk to near neutral to promote the growth of proteolytic organisms (aerobic or anaerobic) which hydrolyze milk protein (casein). The ecological succession continues until the milk is decomposed or stabilized to the point where it is no longer able to support the growth of heterotrophs.

High quality milk has a low bacterial count and contains no pathogenic bacteria. It should have good nutritional composition, flavor, adequate keeping quality, and be free from toxic substances, such as DDT and nitrates. The grades of milk, A, B, C, and Certified,* are classified on the basis of the number of bacteria they contain:

	Raw Milk (ml)	**Pasteurized Milk (ml)**	
Grade	*Standard Plate Count Not to Exceed*	*Standard Plate Count Not to Exceed*	*Coliform Bacteria Not to Exceed*
A	200,000	30,000	10
B	1,000,000	50,000	10
C	>1,000,000	50,000	>10
Certified	10,000	500	1

The *standard plate count* is the official method of enumerating bacteria in pasteurized milk and is often used in the analysis of raw milk. The suitably diluted milk samples are pour-plated with tryptone glucose extract agar (see

*Certified milk is defined by the American Association of Medical Milk Commissions, Inc., as the product of dairies operated in accordance with "Methods and Standards for the Production of Certified Milk" and under direct supervision of local Medical Milk Commissions or other recognized agencies approved by the Association.

Materials section for composition) and counts per ml are made after incubating for 48 hr at 35C.

The second method of counting the number of bacteria in milk is the *direct microscopic count.* A special pipette (Breed) or wire loop which holds 0.01 ml of milk is used to prepare a smear having an area of 1 cm^2 on a special slide or on a clear microscope slide, using a guide card which has several blocks of 1 cm^2 (see Fig. 30-1). The smears are stained with aqueous methylene blue dye. The number of bacteria observed per field is translated into the number of bacteria/ml of the milk sample. The method has some advantages over the standard plate count; for example, it is more rapid, the results being obtained in a few minutes and, in addition, the slide may be kept as a permanent record.

A third and frequently used method for grading milk is the *methylene blue reduction test.* Methylene blue thiocyanate is blue colored in the oxidized state but it is an excellent hydrogen acceptor and becomes colorless in the reduced state. Therefore, the greater the number of bacteria present in an incubated, closed milk sample the more rapidly the dissolved O_2 will be depleted, the oxidation-reduction potential (Eh) lowered, and the shorter the time required to reduce the dye. Hence, Grade A milk requires a longer time for reduction of dye than Grade B, and Grade B milk requires a longer time for reduction of dye than Grade C. The reduction end point is 80 per cent decolorization of the dye. As a rule, Grade A raw milk to be pasteurized should not reduce methylene blue in less than five hours.

For purposes of characterization and classification, skim milk is sometimes used as a culture medium for determining acid production, proteolysis, and to some extent the production of the enzyme rennet. An indicator, litmus, is frequently added to the milk because it indicates reduction (lavender color fades to white) as well as pH changes. Neutral litmus milk has a lavender color, which changes to pink or red with acid production, or to blue if it becomes alkaline.

In this experiment the object is to show you (1) the kind of changes which take place in sterile litmus milk inoculated with certain kinds of organisms; (2) chemical changes produced in raw skim milk (lactic acid production), and (3) some of the tests used in grading raw and pasteurized milk. The experiment will be performed by you and your partner.

PERIOD I

PART A. CHANGES PRODUCED BY PURE CULTURES OF BACTERIA IN LITMUS MILK

MATERIALS (for a pair of students)

1. 9 tubes of litmus milk.

2. Broth cultures of *Streptococcus lactis* (1), *S. faecalis* var. *liquefaciens* (2), *Enterobacter (Aerobacter) aerogenes* (3), *Bacillus cereus* (4), *Alcaligenes viscolactis* (5), *Torulopsis lactis-condensi* (yeast) (6), your partner's unknown culture (7), and your unknown culture (8).

PROCEDURE

1. Label each tube of litmus milk with the name of a culture provided. Label the ninth tube as control.

2. Inoculate each labeled tube of litmus milk with 0.1 ml of the corresponding pure culture.

3. Mix the inoculum with the milk by rotating the tube between the palms of your hands. Incubate the tubes for 2 to 5 days at 37C.

PART B. CHEMICAL CHANGES OF RAW SKIM MILK CAUSED BY NORMAL FLORA (LACTIC ACID MEASUREMENT)

MATERIALS

1. 60 ml sample of good *or* poor quality raw skim milk, whichever assigned.
2. Titration apparatus:
   a. 50 ml burette, clamp, stand.
   b. Phenolphthalein indicator solution.
   c. 1 clean 125 ml Erlenmeyer flask.
   d. Stock solution of 0.10N NaOH.

PROCEDURE

1. Pipette 5.0 ml of the milk into a clean 125 ml Erlenmeyer flask provided.

2. With a pipette, add 5.0 ml distilled water to the 5.0 ml milk sample.

3. Add 3 to 5 drops of phenolphthalein indicator solution to the flask and mix contents thoroughly.

4. Titrate the prepared sample with 0.10N NaOH until the indicator shows a faint pink color which persists for 1 minute. Record the volume of 0.10N NaOH used in the space provided.

PART C. STANDARD PLATE COUNT, DIRECT MICROSCOPIC COUNT, AND METHYLENE BLUE REDUCTION TEST

MATERIALS

1. 1 bottle of sterile, melted, tryptone glucose extract agar (beef extract, 3.0 g; tryptone, 5.0 g; glucose, 1.0 g; agar, 15.0 g in 1000 ml distilled water).

2. 2 sterile dilution water blanks (99 ml/bottle).

3. 6 sterile Petri dishes.

4. A tube of 1:30,000 aqueous solution of methylene blue thiocyanate.

5. Milk samples from Part B for standard plate count.

6. 10 ml of good or poor quality milk in a screw-capped test tube.

7. A canister of pipettes.

8. 3×5 cards.

PROCEDURE

The Standard Plate Count

1. Remove 1.0 ml of raw skim milk from the sample assigned to you in Part B and transfer the milk to a 99 ml dilution blank. Mix thoroughly and label the bottle 10^{-2}.

2. Pipette 1.0 ml of diluted milk sample (10^{-2}) into a 99 ml water blank. Mix thoroughly and label the bottle 10^{-4}.

3. Transfer 0.1 ml of diluted milk from the 10^{-2} dilution bottle into each of 2 Petri dishes and label the dishes as 10^{-3}. Likewise, transfer 1.0 ml from the 10^{-4} bottle and label these dishes as 10^{-4}; transfer 0.1 ml from 10^{-4} bottle into each of the last 2 dishes and label as 10^{-5}.

4. Obtain melted tryptone glucose extract agar from a water bath (45C). Pour approximately 15 to 20 ml of the agar into each dish. As each dish is poured, mix the agar with the diluted milk by rotating and tilting the dish back and forth gently.

5. Allow to solidify, invert the plates, and incubate for 2 to 5 days at 37C.

Direct Microscopic Count

1. Prepare 1 cm² guides on a 3×5 card as directed by your instructor.

2. Place a clean glass slide over the 1 cm² guide on the card (see Fig. 30-2).

3. Mix the milk sample used in Part B by gently rotating, transfer a loopful (0.01 ml) of milk onto the slide over the 1 cm² grid, and spread the sample uniformly over the entire area (1 cm²). Air dry the slide.

4. Flood the dried smear with xylol for about 1 minute and drain; air dry.

5. Flood the dried smear with 95 per cent alcohol for 1 minute; drain and air dry. Steps 4 and 5 are to remove lipids in the milk which may interfere with the staining procedure.

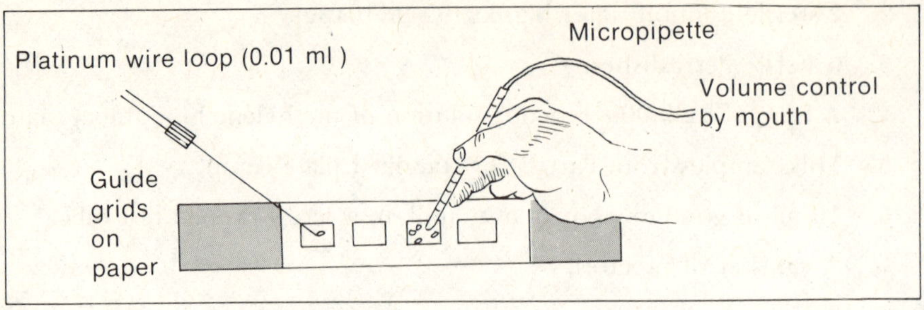

FIGURE 30–2. Direct microscopic count: Milk smears are prepared by use of wire loop or Breed pipette on a glass slide over a guide card.

6. Flood the dried smear with aqueous methylene blue for 3 minutes. Drain off the stain and wash briefly in a tumbler of water. Drain off all the water and blot dry. Label the slide and store it until next period.

Methylene Blue Reduction Test

1. Add 1.0 ml of methylene blue thiocyanate solution into whole milk (10.0 ml) provided in a special screw-capped test tube (total capacity 11 ml).

2. Mix the milk and the dye thoroughly by rapidly rotating the tightly-capped tube between the palms of your hands to insure uniform distribution of dye, cream, and milk. Note: there is no air space in the tube; the presence of oxygen interferes with methylene blue reduction time.

3. Place the tube in a water bath (37C) and record the starting time in the space provided. Examine the tube at the end of 2 hr and thereafter your instructor will check the methylene blue reduction time. (Make certain your tube is well labeled.)

PERIOD II

PROCEDURE

PART A. CHANGES PRODUCED BY PURE CULTURES OF BACTERIA IN LITMUS MILK

1. Examine the changes occurring in litmus milk inoculated with various pure cultures of bacteria. Note the following changes:

a. Presence of bubbles indicates gas formation (G).

b. Change in color of litmus to red indicates acid production (A).

c. Solidifying of milk indicates curdling (C).

d. If a portion of milk becomes clear and watery, this indicates hydrolysis of casein or proteolysis (P).

e. Change in color of litmus to darker blue indicates alkali production (K).

f. When touched with a wire loop, milk is stringy or ropy (S).

g. Change in litmus color to colorless compound indicates reduction (R).

2. Using the letters in parentheses to indicate reactions, record your observations in the space provided.

PART B. CHEMICAL CHANGES OF RAW SKIM MILK CAUSED BY NORMAL FLORA (LACTIC ACID MEASUREMENT)

1. Using a loopful of decomposing milk, prepare a methylene blue stained smear. The smear should be treated with xylol and alcohol as in Period I, Part C, Direct Microscopic Count. Observe the slide with the oil immersion lens and sketch a representative field in the space provided.

2. Measure the amount of titratable acid produced in skim milk (see Period I for the method).

3. Record the volume of 0.10N NaOH used in the space provided.

4. Calculate the acidity in terms of per cent lactic acid by using the formula given below:

$$\%\ \text{lactic acid} = \frac{\text{ml NaOH used} \times 0.009 \times 100}{\text{ml of milk used*}}$$

Record the per cent lactic acid produced in the space provided. Your instructor will explain the derivation of the formula.

PART C. STANDARD PLATE COUNT, DIRECT MICROSCOPIC COUNT, AND METHYLENE BLUE REDUCTION TEST

Standard Plate Count

1. Compute the number of viable organisms per ml of the original milk sample. Use the average from each pair of plates and count only plates where colony numbers are between 30 and 300 per plate.

2. Record your results in the space provided.

Direct Microscopic Count

1. Observe the methylene blue stained slide prepared in the previous period under the oil immersion lens.

---

*Do not include 5.0 ml of distilled water added.

2. Count the number of cells (per field) and repeat the process at least 30 times with a new field each time. Obtain the average number of cells per field by dividing the total number of cells by the number of fields you have observed.

3. Compute the number of bacteria per ml by multiplying the average number of bacteria per field by 500,000 (conversion factor). Your instructor will explain the derivation of the conversion factor.

Methylene Blue Reduction Test

1. Obtain methylene blue reduction time of your milk sample from your instructor and record the time in the space provided. The following table shows the relative quality of raw milk.

Reduction Time	Approx. No. of Bacteria/ml	Quality
2 to 3 hr	<5,000,000	poor milk
4 to 6 hr	<750,000	fair milk
7 to 9 hr	<200,000	good milk
>10 hr	<50,000	excellent milk

DISCUSSION

1. What caused litmus (indicator) to fade to white when certain organisms were cultured in litmus milk?

2. What is the reason for using a nutritionally rich medium such as tryptone glucose extract agar for enumeration of viable bacteria in milk?

3. What is a major disadvantage of using direct microscopic count to enumerate the number of bacteria in Grade A raw milk?

4. What is meant by clump count? Refer to your text, if necessary.

EXPERIMENT 31

Characterization of a Plant Pathogen

It is impossible to estimate the adverse effects on the national economy from crop failures caused by diseases of plants. Throughout history, plant diseases have caused famine, malnutrition, and loss of farms, homes, and the independence of farmers. The potato blight (fungus, *Phytophthora infestans)* in Ireland in the mid-nineteenth century, for example, destroyed most of the potatoes within a two-year period.

Plant infections, like those of man and other animals, are caused by viruses, bacteria, and fungi. Among the bacteria that are pathogenic for plants are species of *Pseudomonas,* which cause leaf spot, stripe, and wilt of tomato and other plants; species of *Xanthomonas,* slime-producing bacteria which cause wilt of sweet corn and others; species of *Agrobacterium,* which produce hypertrophies, called crown-gall disease (or plant cancer), of tomato and various other plants; species of *Corynebacterium,* which cause a vascular disease of alfalfa, ring spot of potato, and many other plant diseases. (It should be remembered that species of *Pseudomonas* and *Corynebacterium* include animal as well as plant parasites and pathogens.)

Species of *Erwinia* are capable of invading the tissues of living plants to produce soft rot in potato, carrot, onion, and cabbage, "fire blight" (cankers) of apple, dry necrosis, galls, and wilt in tobacco, tomato, potato, and other plants. Species of *Erwinia* (*E. carotovora* and *E. atroseptica*) are gram-negative, small (0.7×1 to $2\ \mu m$) motile rods. The bacteria produce an extracellular enzyme, middle lamella pectinase, which dissolves the middle lamella (cell cementing agent consisting primarily of pectins) and loosens the cells, which in turn facilitates the invasion of the surrounding tissues. Infected potato tubers may become softened and, as the disease advances, become watery or slimy.

Many plant diseases can be controlled by selection of plants resistant to disease (while there is no variety of plant completely resistant to disease, some are more resistant than others; resistant strains may be obtained by selective breeding), proper treatment of the seed, prevention of bruising of plant parts (stems, leaves, fruit, etc.) during harvest and storage, and storage at low temperature.

These are only a few examples of plant disease caused by bacteria and, being an introductory bacteriology course, we will limit our experiment to the stem rot of potato caused by *Erwina atroseptica,* to soft rot without discoloration caused by *Erwinia carotovora,* and to a demonstration of crown-gall production in the tomato plant by species of *Agrobacterium.*

PERIOD I

PART A. SOFT ROT OF POTATO

MATERIALS (for a pair of students)

1. Small, sound, firm potato.
2. Tumbler with 95% ethyl alcohol.
3. Tumbler with 1:1000 aqueous $HgCl_2$ solution.
4. Several sterile paper towels.
5. 1 sterile paring knife.
6. 1 test tube brush.
7. 3 sterile Petri dishes with filter paper.
8. Broth cultures of *Erwinia carotovora* and *E. atroseptica.*
9. 1 bottle of sterile distilled water.
10. 1 tube of sterile distilled water (used as inoculum for control).

PROCEDURE

1. Aseptically pipette about 6.0 ml of sterile water into each of 3 Petri dishes (this is to moisten the filter paper to prevent dehydration of inoculated potato).

2. Scrub the potato thoroughly with the brush under running cold tap water. (Make sure that the pores and crevices of the potato are also scrubbed.)

3. After the thorough scrubbing, place the potato in 95 per cent ethyl alcohol for 2 minutes to drive air from the pores and crevices.

4. Remove the potato from the alcohol and immerse it in the $HgCl_2$ solution for 10 minutes. (This procedure is an attempt to minimize the numbers of extraneous microorganisms on the exterior of the potato.)

5. Remove the potato from the solution with flame-sterilized forceps and place it between the sterile paper towels provided. Dry the potato by gently pressing with the paper towels.

6. Lift the paper towel slightly and cut 3 slices of potato each approximately 1 cm thick, using the sterile knife. (Note that the paper towel serves as a roof to prevent contamination of the potato from the air and the desk.)

7. With sterile forceps, aseptically place one slice of potato into each of 3 Petri dishes. Carefully carve out, with the sterile knife, a small well in the center of the potato to receive the inoculum. Your instructor will demonstrate the procedure. Label one dish as E.c., one as E.a., and the third dish as control.

8. Inoculate the potato slice labeled E.c. with 1 to 2 drops of *E. carotovora* in the well. Inoculate the slice in the dish labeled E.a. with *E. atroseptica,* and add 1 to 2 drops of sterile distilled water to the control.

9. Incubate the dishes in your desk drawer, right side up, until evidence of rotting is observed (usually within 2 to 5 days).

10. Prepare a gram-stained smear of pure cultures of *E. carotovora* and *E. atroseptica.* Observe the smears under the oil immersion objective. Record the gram reaction and sketch the cells from representative fields in the space provided.

PART B. DEMONSTRATION OF GALL PRODUCTION BY AGROBACTERIUM TUMEFACIENS ON TOMATO STEM

MATERIALS

1. Healthy potted tomato plants 6 to 8 weeks old.
2. Broth culture of *Agrobacterium tumefaciens* 48 hr old.
3. 2 sterile hypodermic syringes.
4. 1 tryptic soy agar plate.
5. 1 sterile distilled water blank.

PROCEDURE

Your instructor will perform the following demonstration. You are to observe each step of the demonstration experiment.

1. The instructor will prepare a streak plate of *Agrobacterium tumefaciens* on tryptic soy agar. The plate will be incubated for 48 hr at room temperature.

2. Your instructor will fill a 1 ml sterile syringe with a broth culture of *A. tumefaciens* and will inoculate the tomato plant by puncturing the tissue near the stem node (the node is where the petiole of the leaf originates from the stem) and injecting a small amount of the culture. He will repeat this process at 2 to 3 different locations. At this point make a brief sketch of the tomato plant and mark the sites where he has injected the culture.

3. A second plant will be inoculated as a control using a second syringe filled with sterile distilled water. Again, make a sketch and record the locations where water was injected.

4. The inoculated plants will be incubated for 2 to 3 weeks in a special area of a greenhouse to provide suitable light, temperature, and humidity. At the end of the incubation period the plants will be examined for symptoms of the disease (tumors).

PERIOD II

PART A. SOFT ROT OF POTATO

MATERIALS

1. 2 plates of sterile yeast extract–mannitol agar (see Experiment 26, Part A, 2, for formula).

PROCEDURE

1. Examine the inoculated slices of potato and in the space provided record your observations in terms of change in color, odor, and structure of the tissues.

2. Compare the changes caused by *E. carotovora* and *E. atroseptica* with the control and record your observations in the space provided.

3. Prepare a gram-stained smear from the decaying areas of both slices of potato. (You may put both cultures on the same slide provided smears do not contact each other.) Observe the smears under the oil immersion objective and record the gram reaction. Sketch cells from representative fields. Are there any differences in gram reaction? In cell morphology?

4. Prepare streak plates of *E. carotovora* and *E. atroseptica* on yeast extract–mannitol agar plates provided. Invert and incubate the plates for 48 hr at 30C.

PART B. DEMONSTRATION OF GALL PRODUCTION

MATERIALS

1. Pure culture of *A. tumefaciens* on agar slant.

PROCEDURE

1. Observe the inoculated plants and the tumors or galls produced on the inoculated areas of the stem. Compare the infected plant with the control plant. Record your results in the space provided.

2. Your instructor will macerate tumor tissue for Gram stain and the tissue will be provided to you along with a pure culture of *A. tumefaciens*.

3. Prepare smear for Gram stain as follows:

   a. Prepare a smear of pure culture of *A. tumefaciens* on a slide, slightly off center. Place a loopful of macerated tumor tissue adjacent to the pure culture. Smears should not be in contact with each other. Air dry, heat fix, and stain with Gram's stain.

b. Observe the stained smears with the oil immersion objective and compare the cell morphology observed in the tumor tissue with that of the pure culture of *A. tumefaciens.* Record your observations in the space provided.

PERIOD III

PART A. SOFT ROT OF POTATO

PROCEDURE

1. Examine the streak plates and describe the characteristics of the colonies which developed. Record your observations in the space provided.

2. Prepare Gram stained smears of *E. carotovora* and *E. atroseptica* from characteristic colonies found on the dishes. Examine the smears under the oil immersion objective and compare with your observations of the original pure cultures and smears prepared from rotted material. Record your observations in the space provided.

RESULTS

PART A. SOFT ROT OF POTATO

1. Sketch representative microscopic fields of Gram stained preparations of the following:

a. Pure cultures of *E. carotovora* and *E. atroseptica.*

E. carotovora *E. atroseptica*

b. Slides prepared from rotted materials.

E. carotovora *E. atroseptica*

c. Slides prepared from the colonies formed on the streak plates.

E. carotovora *E. atroseptica*

2. Briefly discuss the differences (morphological, staining reaction of the cells, etc.) among the slides prepared from the three different sources.

3. Give word descriptions of characteristic rotting of the potato by the two cultures used.

PART B. DEMONSTRATION OF GALL PRODUCTION

1. Sketch tomato plants and indicate sites of injection.

Inoculated plant

Control plant

2. Sketch infected plant and show any changes which occurred at inoculated site.

3. Sketch representative microscopic fields of Gram stained preparations of the following:

Pure culture of *A. tumefaciens*

Slide prepared from tumor tissue

DISCUSSION

1. What are Koch's postulates? As this experiment was performed, did it establish the causative agent of infection? Describe briefly.

2. Both *E. carotovora* and *E. atroseptica* are unable to hydrolyze starch and cellulose but are able to cause disease in plants. How do these organisms incite disease in potato?

EXPERIMENT 32

Medical Bacteriology: Isolation and Identification of Certain Pathogenic Cocci

Among the most frequently encountered pathogenic bacteria are the gram-positive, pyogenic or "pus-producing" cocci. These cocci commonly include species of *Staphylococcus* (family Micrococcaceae) and species of *Streptococcus* (family Streptococcaceae). (See *Bergey's Manual of Determinative Bacteriology*, 8th ed., Part XIV.) These organisms can be isolated from patients as well as from healthy carriers. It has been reported that as high as 80 to 90 per cent of hospital personnel are carriers of staphylococci (10 to 75 per cent also carry pyogenic streptococci). Many of the staphylococcal isolates are resistant to one or more types of penicillin. Other pathogenic cocci of public health concern are gram-negative, strictly aerobic organisms in the family Neisseriaceae: *Neisseria gonorrhea,* the cause of veneral disease in man, and *Moraxiella,* cocco-bacilli frequently isolated from human conjunctivitis.

Although preparing a Gram stain of an isolate is helpful in identifying cocci, it is not sufficient to differentiate pathogenic from saprophytic types or to identify the species. For example, the family Micrococcaceae contains both the genus *Staphylococcus,* regarded as pathogenic for man and animals, and the genus *Micrococcus,* regarded as a nonpathogenic saprophyte. Both types of bacteria are gram-positive, spherical in shape, occur in irregular clumps, and are often found mixed together under natural conditions. Consequently, in order to differentiate between the two genera, the bacteria must be isolated in pure culture and subjected to various cultural and physiologic tests. Virulence among the staphylococci is highly variable. *S. epidermidis* generally lacks virulence and does not produce α-hemolysin or coagulase (a factor promoting the coagulation of citrated plasma); *S. aureus* isolated from infected individuals almost always produces α-hemolysin and coagulase and ranges from low to high virulence depending on the strain. See your text for further details.

Among the tests frequently used in differentiating *S. aureus* from species of *Micrococcus* are the ability to ferment mannitol, the ability to utilize inorganic nitrogen ($NH_4H_2PO_4$) as a sole source of nitrogen, and the growth characteristics on blood agar. As a rule, *S. aureus* is unable to utilize $NH_4H_2PO_4$ as a nitrogen source, but many species of the nonpathogenic micrococci are able to do so. Most species of micrococci are aerobic and unable to ferment mannitol, but

S. aureus is facultative, able to ferment mannitol with the production of acid, and will also tolerate a high salt (7.5% NaCl) concentration.

Growth characteristics and hemolytic activities on blood agar are often used in differentiating pyogenic cocci—staphylococci and streptococci. The term "hemolytic activities" refers to the destruction of red blood cells by hemolysins released by colonies growing on blood agar plates. The hemolytic patterns can be divided into three major types: (1) complete hemolysis, (2) incomplete hemolysis, and (3) nonhemolysis. These have traditionally been denoted by Greek letters, α, β, γ, δ, and so forth, but without any attempt to standardize the nomenclature applied to different genera of bacteria. Thus, the α-hemolysin of staphylococci produces a hemolytic pattern completely different from that of the α-hemolysin of streptococci. Table 32–1 is a simplified and abbreviated comparison of the hemolysins of these two genera.

TABLE 32–1
BLOOD AGAR TYPES* (SHEEP BLOOD)

Hemolytic Pattern Observed	Designation for *Staphylococcus*	Designation for *Streptococcus*
1. *Complete hemolysis:* complete destruction of erythrocytes and hemoglobin	α-(hemolyzed zone is wide) or δ-(zone is narrow) hemolysis	β-hemolysis
2. *Incomplete hemolysis:* partial destruction of erythrocytes and hemoglobin; only partial clearing around colony	β-hemolysis or hot-cold hemolysis†	α-hemolysis: partially destroyed erythrocytes and hemoglobin having a green or brownish-green color; hence the group is often called viridans.
3. *No hemolysis:* growth occurs on agar but shows no visible change in erythrocytes	No specific designation	γ-hemolysin (nonhemolytic type)
4. Combinations: more than one kind of hemolysis by the colonies on blood agar		

*Nomenclature of hemolysins, especially of *Staphylococcus,* is based on the kind of susceptible erythrocytes (sheep, horse, rabbit, guinea pig, or human blood), the hemolytic pattern observed, and cross-neutralization patterns with antisera.

†Plates showing partial hemolysis will advance to complete lysis if they are held overnight in the refrigerator after being removed from the incubator. Mechanism poorly understood.

It must be emphasized that as a basis for classifying bacteria, e.g., *Staphylococcus* and *Streptococcus,* hemolytic reactions are far from satisfactory because not all pathogenic species or strains are hemolytic; in addition, many saprophytes are capable of complete hemolysis of the blood agar. Consequently, hemolysis by itself does not indicate pathogenicity of the organism. The classification system for hemolytic reactions is incomplete and confusing to the neophyte, but it has become so firmly established that it cannot be ignored.

The classification of pyogenic and nonpyogenic streptococci is just as complex as that of the staphylococci and micrococci. The organisms are frequently classified by serologic means. The student is encouraged to read the textbook concerning the immunologic grouping of β-hemolytic streptococci by Lancefield.

The purposes of this experiment are to attempt to isolate pure cultures of staphylococci and streptococci from blood agar plates prepared from the nose and throat of an infected person, and to demonstrate some of the differences between these two organisms by using simple cultural and physiologic tests. The experiment is to be performed by you and your partner.

PERIOD I

MATERIALS (for a pair of students)

1. Several blood agar plates pre-prepared from the nose and throat of an infected person.

2. 7 blood agar plates (see Appendix for formula).

3. 3 mEnterococcus agar plates (see Appendix for formula).

4. Pure cultures of *Staphylococcus aureus, Micrococcus conglomeratus* or *M. albus, Streptococcus pyogenes, S. faecalis,* and *Bacillus cereus;* your unknown, your partner's unknown.

PROCEDURE

1. Label the blood agar plates as follows: 1, throat or nose + your initials; 2, throat or nose + your partner's initials; 3, divide bottom of the plate in half with marking pencil and label one side as S.a. and the other as M.a. or M.c., whichever is assigned; 4, divide bottom of plate in half and label one side S.p., the other S.f.; 5, B.c.; 6, your unknown isolate; 7, your partner's unknown isolate.

2. With a marking pencil, label one mEnterococcus plate as nose and throat culture.

3. Divide the 2 remaining mEnterococcus agar plates into 4 equal sectors each and label with the names of the pure cultures provided, your partner's unknown, your unknown, and the last sector as control.

4. With a sterile needle, pick a colony resembling staphylococci from the pre-prepared blood agar plate. Prepare a streak plate on the appropriate blood agar plate (follow the streaking procedure you have used throughout this semester).

5. Your partner will follow the procedure in Step 4, except he will select a whitish, nonhemolytic colony resembling micrococci.

6. Either you or your partner will select a very minute colony with a hemolytic zone (most resembling a streptococcus colony; if you are uncertain, consult your instructor). Follow the procedure in Step 4 and prepare a streak plate on the appropriately labeled mEnterococcus agar plate.

7. Using a flame-sterilized loop, prepare blood agar streak plates of the pure cultures provided, of your unknown, and of your partner's unknown on the appropriately labeled plates from Step 1.

8. Follow the same procedure using the mEnterococcus agar plates prepared in Steps 2 and 3.

9. Invert the plates and incubate for 24 to 48 hr at 37C. Record the type of hemolysis. The blood agar plates will then be incubated for 24 hr at 4C to allow any "hot-cold" hemolytic patterns to develop.

PERIOD II

MATERIALS

1. 6 tubes of mannitol salt agar (Difco). See Appendix for formula. The medium contains phenol red which serves as pH indicator. The color of the indicator at pH 7.5 ± 0.1 is orange; below 7.4 (more acid) the color becomes yellow, and above 7.6 it becomes red.

2. 6 tubes of glucose ammonium acid phosphate agar. See Appendix for formula. The medium contains brom cresol purple as pH indicator. The indicator is purple at pH 6.9 ± 0.1; at less than 6.6 it becomes yellow, and above 7.2 the indicator becomes blue.

PROCEDURE

1. Examine the blood agar plates (24 hr at 37C and 24 hr at 4C) and describe:

   a. Whether the throat or nose culture and the species tested caused hemolysis or failed to produce hemolysis on the agar plates. Record your observations in the space provided.

   b. If any pure culture tested produced hemolysis, describe the type of hemolysis produced (refer to introductory statement if necessary).

   c. Select 2 colonies resembling hemolytic *Staphylococcus* (see pure culture streak plate for comparison) and 2 resembling non-hemolytic *Micrococcus* sp from the plates streaked with nose or throat cultures. Mark these colonies with circles and number the circles. (Before selecting the colonies, you should become thoroughly familiar with the colonial morphology of each culture by closely observing the pure culture on the blood agar plate.) If you have difficulty in deciding, call your instructor for help.

   d. Prepare a Gram stained smear from each of the 2 colonies selected resembling hemolytic *Staphylococcus* sp and from a colony of *S. aureus*. You may put all 3 smears on one glass slide provided you label the slide and the smears do not touch each other. Do the same for *Micrococcus* sp.

e. Select one smear which resembles most closely the pure culture of *Staphylococcus* and mark the corresponding colony with X. Do the same with the *Micrococcus* slide. Record your results in the space provided.

2. Label the mannitol salt agar slants as follows: one slant as S.a., one as M.a. or M.c. (whichever assigned), one as the selected *Staphylococcus* sp from nose or throat, one as selected *Micrococcus* sp from nose or throat, one as your unknown, and the last as your partner's unknown.

3. Similarly label the glucose ammonium acid phosphate agar slants.

4. Inoculate the agar slants (both media) labeled S.a. by removing a small amount of the growth from a colony of pure culture of *S. aureus*. Each tube should be inoculated in two ways: first, make a stroke on the surface of the slant, then stab the loop through the entire column of agar. The stab inoculation is to determine the organism's ability to grow in the absence of air. Do the same with the pure culture of *M. albus* or *M. conglomeratus* and with the colonies you and your partner selected (marked with X) as most likely to be hemolytic *Staphylococcus* sp and nonhemolytic *Micrococcus* sp.

5. Incubate the inoculated tubes for 48 hr at 37C.

6. From the blood agar plates, make Gram stained preparations from colonies of *S. pyogenes* and *S. faecalis*. Observe the smears under the oil immersion objective. Compare the cell morphology, arrangement, and gram reaction with the smears of the pure cultures of staphylococci and micrococci prepared above. Note the distinctly different cell arrangement produced by streptococci (cells in chains, like strings of pearls).

7. Examine the mEnterococcus agar plates and describe whether or not the organisms inoculated showed growth. The medium is selective for Group D streptococci (fecal streptococci of Lancefield immunologic groups). Fecal streptococci will not only grow on this medium but will produce characteristic very small, reddish-brown colonies. Record your results in the space provided.

PERIOD III

MATERIALS

1. Broth cultures of *S. aureus* and *S. epidermidis*.
2. Citrated rabbit plasma or commercial coagulase plasma EDTA (Difco).

PROCEDURE

1. Examine each culture on mannitol salt agar slants. Answer the following questions and record your observations in the space provided.

a. Did growth occur on the surface of the slants?

b. Did growth occur in the stab?

c. Was there any change in the phenol red indicator? Acidic or alkaline change?

Remember, growth on the medium indicates that the organism is salt tolerant (the medium contains 7.5% NaCl) and if the growth occurred throughout the stab, this means the organism is facultative. If the indicator has changed to yellow color, then the culture is capable of fermenting mannitol to acids. Pathogenic *S. aureus* will tolerate the salt concentration as well as ferment mannitol with the production of acids.

2. Examine each culture on glucose ammonium acid phosphate agar slants. Answer the following questions and record your observations in the space provided.

a. Did growth occur on the surface of the slant?

b. Did growth occur in the stab?

c. Was there any change in BCP indicator? Acidic or alkaline? As a rule, saprophytic species of *Micrococcus* are capable of utilizing inorganic nitrogen (as in $NH_4H_2PO_4$) as a sole source of nitrogen, but pathogenic *S. aureus* is unable to use inorganic nitrogen.

d. Your instructor will demonstrate the coagulase test. Provided the class is not too large, those students who isolated what they believe to be *S. aureus* from nose or throat may wish to perform the coagulase test on the isolate. Inoculate a tube of BHI broth with the isolate and incubate overnight at 37C. The broth can be used the following day to run the test. (0.3 ml of broth culture is added to 0.5 ml of freshly prepared citrated rabbit plasma diluted 1:5 in normal saline. Incubate tube at 37C and observe every 15 minutes for clotting over a 2 hr period; alternatively, use commercial coagulase plasma* and follow directions of manufacturer.)

*EDTA is preferable as an anticoagulant over citrate, because some bacteria utilize the citrate; when the citrate is utilized, the calcium of the plasma is no longer chelated and the plasma clots, thus giving a false positive coagulase reaction.

RESULTS

1. Fill in the table below with your observations of the cultures on the blood agar plates.

Culture	Type of Hemolysis	Gram Reaction	Word Description of Colonies on Blood Agar Plates
S. aureus			
M. albus or *M. conglomeratus*			
S. pyogenes			
S. faecalis			
B. cereus			
S. aureus type from nose or throat			

(Result 1 continued on following page)

(Result 1 continued)

Culture	Type of Hemolysis	Gram Reaction	Word Description of Colonies on Blood Agar Plates
Micrococcus type from nose or throat			
Your unknown isolate			

2. Fill in the table below with description of growth on mEnterococcus agar.

Culture	Description of Growth
S. aureus	
M. albus or *M. conglomeratus*	
S. pyogenes	
S. faecalis	
B. cereus	
S. aureus type from nose or throat	
Micrococcus type from nose or throat	
Your unknown isolate	

3. Complete the table below with your observations on (1) mannitol salt agar slants and (2) glucose ammonium acid phosphate agar slants. Use + to indicate growth and − to indicate absence of growth.

Cultures		Growth *Surface*	Growth *Stab*	Change in Color of the Indicator	Word Description of the Growth
S. aureus	1				
	2				
M. albus or *M. conglomeratus*	1				
	2				
S. aureus type from nose or throat	1				
	2				
Micrococcus type from nose or throat	1				
	2				
Your unknown isolate	1				
	2				

4. Optional: If *S. aureus* was isolated from nose, was the test for coagulase production negative or positive?
What is the significance of this finding?

DISCUSSION

1. How would you distinguish hemolytic staphylococci from hemolytic streptococci? List a few methods.

2. List some of the diseases produced by streptococci (refer to your text, if necessary).

3. Are any of the streptococci beneficial to man? In what way(s)?

4. Other than the nose or throat, where else might you find staphylococci on the body?

EXPERIMENT 33

Medical Bacteriology: Identification of "Enteric Bacteria"

The terms "enterics" and "enteric bacteria" refer to the gram-negative, facultatively aerobic, nonsporeforming rods of the family Enterobacteriaceae. The organisms in the family are common inhabitants of the colon and intestinal tract of man and other animals. Some of the enterics are commensals; some are symbionts; others are clearly pathogens; and some are regarded as "opportunists," that is, they normally do not cause disease unless introduced into the body by an unnatural route, or the normal protective flora of the intestinal tract have been disturbed or eliminated (by antibiotic therapy), or the host becomes debilitated (by cancer therapy, malnutrition, etc.). Species of *Proteus,* frequently found in normal feces, have been isolated from patients with an infection of the urinary tract; certain strains of *Escherichia coli* are known to cause infantile diarrhea, frequently fatal. Of much greater medical concern, however, are species of *Salmonella* and *Shigella,* the causative agents of enteric fever, food infection, and bacillary dysentery.

Isolation of species of *Salmonella* and *Shigella* from feces is complicated and time-consuming because the colon contains a vast array of morphologically and physiologically similar bacteria belonging to several genera. The rapid isolation and precise identification of an enteric pathogen having many close relatives presents a real challenge to the diagnostic bacteriologist. For example, isolation of *Salmonella* or *Shigella* requires the use of selective (enrichment) media to favor the growth of these organisms while discouraging the growth of bacteria such as coliforms. Enrichment media, such as selenite broth or tetrathionate broth, are frequently necessary for the detection of these pathogens in stool specimens when they are present in small numbers. Subcultures are usually made on selective media such as bismuth sulfite (BS) agar, brilliant green (BG) agar, or *Salmonella-Shigella* (SS) agar. The isolates from these media are separated into genera or groups (depending on the classification scheme) with such media as lysine-iron-agar (LIA) or triple sugar-iron-agar (TSI), followed by use of commercially prepared media such as Enterotube* (a compartmentalized tube capable of testing glucose and lactose fermentation, production of H_2S, and utilization of lysine, ornithine, indole, dulcitol, phenylalanine, urea, and citrate). Organisms so identified are further classified by serologic study in order

*Roche Diagnostics, Division of Hoffman-LaRoche Inc., Nutley, N.J.

TABLE 33-1. REPRESENTATIVE ANTIGEN SCHEME FOR *SALMONELLA*.

Type	Assigned Serologic Group	O* Antigens	H Antigens	
			Phase 1	*Phase 2*
(Monophasic) *S. typhi*	D	9, **12** Vi :	d	
(Diphasic) *S. paratyphi B*	B	1, **4,** 5, 12 :	b –	1, 2
(Monophasic) *S. montevideo*	C_1	6, **7** :	g, m, s	
(Nonmotile) *S. pullorum*	D	9, 12 :		

*Boldface antigen is the identifying group antigen.

From *BBL Manual of Products and Laboratory Procedures,* 5th ed., 1968. Division of Becton, Dickinson and Company, Cockeysville, Maryland 21030.

to distinguish one strain from another of the same species. *Salmonella* serotypes are characterized by the identification of both somatic "O" and flagellar "H" antigens. A thermolabile envelope antigen, known as Vi antigen, is frequently found in freshly isolated strains of *S. typhi* and *S. paratyphi C* and aids in identification of these types as shown in Table 33-1, the O antigens are designated by Arabic numerals and the H antigens by Arabic numerals and letters. The O antigens are separated from the H antigens by a colon (:) and from phases of diphasic flagellar types by a dash (–). See text for fuller discussion.

The Kauffman-White Schema is usually used for serologic identification of the *Salmonella* species. The groups are designated by Roman capital letters A to H. However, most *Salmonella* sp belong to somatic antigen groups A to E. By the use of polyvalent O antisera and six individual O antisera representing groups A to E, *Salmonella* types isolated in this country can be grouped; the use of six additional sera prepared against flagellar antigens a, b, c, d, i, and 1 to 7 allows exact identification of some of the more important *Salmonella* serotypes causing human disease. More complete identification systems are given in P. R. Edwards and W. H. Ewing, *Identification of Enterobacteriaceae,* 3rd ed., Burgess Publishing Co., 1972.

In diagnostic bacteriology, accurate and rapid identification (not necessarily classification) of the causative agent of the disease is essential for proper and prompt treatment of the patient. There are many methods—often exceedingly sophisticated and "shortcut"—of identifying some of the enteric organisms, both

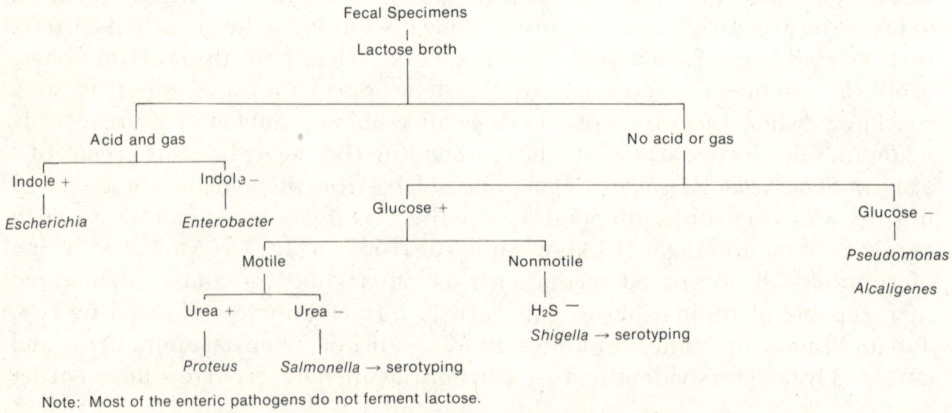

FIGURE 33–1. Differentiation of enteric organisms.

nonpathogenic and pathogenic. The accompanying classic outline, however, is still useful for the beginning bacteriologist (Fig. 33-1).

The purposes of this experiment are (1) to acquaint you with a routine procedure used in clinical and public health microbiology, and (2) to familiarize you with some of the biochemical activities of the pure cultures grown in either lysine-iron-agar or triple sugar-iron-agar (see Appendix for formulas of these media). The media are recommended for use in the identification of gram-negative enteric organisms in the routine examination of stool. A medium such as TSI aids in determining if a pure culture isolate (1) is fermentative or oxidative in its metabolism of carbohydrates, (2) ferments glucose alone, or if it can ferment lactose or sucrose, and (3) produces H_2S by reductive assimilation of sulfur-bearing amino acids, methionine, cystine, or cysteine. The experiment will be performed by you and your partner.

PERIOD I

MATERIALS (for a pair of students)

1. Pure cultures of *Escherichia coli, Proteus vulgaris, Pseudomonas aeruginosa, Salmonella typhimurium,* and *Shigella flexneri.* (Note: 3 of the 5 cultures listed are potentially pathogenic for man. Aseptic technique must be practiced to avoid infection by these organisms.)
2. Your partner's and your unknown isolates.
3. 8 tubes of urea broth (Difco) (see Appendix for formula).
4. 8 tubes of TSI agar or LIA agar (slants).
5. 2 plates of EMB agar.
6. 2 plates of MacConkey agar.
7. 2 plates of SS agar.

PROCEDURE

1. Label the tubes of urea broth with the names of the pure cultures provided and your partner's and your unknowns. The last tube should be used as a control.
2. Inoculate the tubes with one loopful of the cultures, including unknowns.
3. Divide each of the EMB, MacConkey, and SS agar plates into four sectors. Label each sector with the name of the organism to be streaked; include the unknown isolates.
4. Streak the sectored plates with the pure cultures provided. Streaks should be made to form isolated colonies. Use your partner's unknown and your unknown in two of the three remaining sectors. Use the last sector as control.
5. Label LIA or TSI agar slants with the names of the pure cultures

TABLE 33-2
SUMMARY OF TRIPLE SUGAR-IRON-AGAR REACTIONS ON MEMBERS OF THE FAMILY ENTEROBACTERIACEAE*

Organism	Slant	Butt	Gas	H_2S
Escherichia	A (K)	A	+(−)	−
Shigella	K	A	−	−
Salmonella typhi	K	A	−	+(−)
Other *Salmonella*	K	A	+	+ + +(−)
Arizona	K (A)	A	+	+ + +
Citrobacter	K (A)	A	+	+ + +
Edwardsiella	K	A	+	+ + +
Klebsiella	A	A	+ +	−
Enterobacter	A	A	+ +	−
E. hafniae	K	A	+	−
Serratia	K or A	A	−	−
Proteus vulgaris	A (K)	A	+	+ + +
P. mirabilis	K (A)	A	+	+ + +
P. morganii	K	A	−(+)	−
P. rettgeri	K	A	−	−
Providencia	K	A	+ or −	−

*K, alkaline; A, acid.

(From Blair, J. E., Lennette, E. H., and Truant, J. P., eds. *Manual of Clinical Microbiology*. Baltimore: Williams & Wilkins Co., 1970, p. 157. Reprinted by courtesy of the American Society for Microbiology.)

provided, your partner's unknown, your unknown. The last tube should be used as a control.

6. Inoculate the LIA or TSI agar slants with the pure cultures by first stroking on the surface of the slant, then stabbing the entire length of the agar column to promote anaerobic growth of the organisms. Do the same using the unknown isolates.

7. Incubate the agar plates and tubes (including uninoculated controls) for 72 hr at 37C. Either you or your partner should make daily observations of these plates and tubes for the duration of the incubation period. Record any changes taking place in the media in the space provided. Before you make your observations, it is expected that you will have become familiar with the composition of the media and with the kinds of reactions which these organisms promote in TSI or LIA agar (see Tables 33-2 and 33-3).

TABLE 33–3
SUMMARY OF LYSINE-IRON-AGAR REACTIONS ON MEMBERS OF THE FAMILY ENTEROBACTERIACEAE*

Organism	Slant	Butt	Gas	H_2S
Escherichia	K	K or N	− or +	−
Shigella	K	A	−	−
Salmonella	K	K or N	−	+(−)
S. typhi	K	K	−	+ or −
Paratyphi A	K	A	+ or −	− or +
Arizona	K	K or N	−	+(−)
Citrobacter	K	A	− or +	+ or −
Edwardsiella	K	K	− or +	+
Klebsiella	K or N	K or N	+ or −	−
Enterobacter				
cloacae	K or N	A	+ or −	−
aerogenes	K	K or N	+(−)	−
hafniae	K	K or N	− or +	−
Serratia	K or N	K or N	−	−
Proteus				
vulgaris	R	A	−	−(+)
mirabilis	R	A	−	−(+)
morganii	K or R	A	−	−
rettgeri	R	A	−	−
Providencia	R	A	−	−

*K, alkaline; N, neutral; A, acid; R, red (oxidative deamination).

(From Blair, J. E., Lennette. E. H., and Truant. J. P., eds. *Manual of Clinical Microbiology*. Baltimore: Williams & Wilkins Co., 1970, p. 157. Reprinted by courtesy of the American Society for Microbiology.)

PERIOD II

RESULTS

1. Observe urea broth tubes and record your observations in the table (Part 3) below. Red indicates urea hydrolysis.

2. Describe growth characteristics of each organism cultured on EMB, MacConkey, and SS agar.

Organism	Medium		
	EMB Agar	*MacConkey Agar*	*SS Agar*
E. coli			
P. vulgaris			
P. aeruginosa			
S. typhimurium			
S. flexneri			
Your unknown isolate			

3. Observe the reactions on LIA or TSI agar slants and record your observations below. Use symbols A = acid, K = alkaline, N = no change or neutral, R = red; Gas = + or −, H_2S = + (black ppt) or −. Include results of urea hydrolysis in the appropriate column; use R to indicate hydrolysis. (See Tables 33-1 and 33-2).

Organism	Reactions on LIA or TSI Agar Slants				Urea hydrolysis
	Slant	*Butt*	*Gas*	*H_2S*	
E. coli					
P. vulgaris					
P. aeruginosa					
S. typhimurium					
S. flexneri					
Your unknown					

(Adapted from Ewing, W. H. *Differential Reactions of Enterobacteriaceae.* Lecture Outline Series, II. Atlanta: National Communicable Disease Center, 1966.)

DISCUSSION

1. Most of the enteric pathogens do not ferment lactose. What is the purpose of using lactose broth in the diagnosis of the organism? (See flow chart on p. 306.)

2. Determination of the production of the enzyme urease plays an important role in distinguishing *Proteus* from *Salmonella* species. What are the similarities and dissimilarities of species in these two genera?

EXPERIMENT 34

Immunology and Serology

When a foreign substance is introduced into the body, the body defends itself by attempting to destroy the substance. Phagocytic cells engulf the material, if particulate, and attempt to digest it; highly specific antibodies (serum globulins) are synthesized to neutralize or detoxify any toxic material present and to promote phagocytosis by covering particulate matter with antibody (called *opsonization*). Those substances which elicit the formation of antibodies are called antigens. Almost any proteinaceous substance which is foreign to the tissues can act as an antigen; examples include bacterial cells, viruses, fungi, plant extracts, and blood or tissue from another species.

Immunology, the area of science dealing with such phenomena, has traditionally been one of the subjects covered under the broad scope of microbiology, particularly because immunity to many infectious diseases depends on the development of specific humoral (antibody-mediated) immunity and cellular (cell-mediated) immunity, and also because certain microbial diseases can be treated or prevented by the administration of antisera (sera containing antibodies) or vaccines. In recent years, however, immunology has become a field of study in its own right and today we find immunologic techniques being employed by a wide variety of professionals other than microbiologists: biochemists characterizing proteins, surgeons checking on histocompatability before transplantation, geneticists studying immunodeficiencies, blood bank personnel checking blood types, etc. New, highly sophisticated techniques and a greater knowledge of what antibodies are and how they are synthesized has made immunology a truly fascinating part of modern biology.

Serology (the use of serum or its components) has been and will continue to be an important tool for microbiologists. It has many practical applications. For example, in the previous experiment we explained how the serotypes of salmonella may be identified by the types of antigens present on the envelope, cell wall, and flagella. Serotyping is important not only for epidemiologic purposes, but also because certain serotypes may be pathogenic while others are not (e.g., the enteropathogenic *E. coli* generally belong to one of several serotypes).

Widely used serologic reactions include slide and tube agglutinations, flocculation, complement fixation, fluorescent antibody staining, neutralization, precipitation in capillary tubes, immunodiffusion (precipitation in agar or other support materials), and immunoelectrophoresis (electrophoresis followed by immunodiffusion). The type of test selected will depend in part on whether the antigen is soluble or particulate, the amount of reactants available, and the kind of information needed. The objective of all of these reactions, however, is to bring antigen and antibody together in proper proportions and to allow them to combine in some manner that can be visualized or measured. The tests may be of a qualitative nature—determining whether the antigen and antiserum (antibody) are specific for one another—or they may be quantitative—measuring the *amount* of either antibody or antigen present in the reactants. An added complexity of many serologic reactions is that the antigen, such as a suspension of bacterial cells, may contain many different antigenic substances and each substance may have more than one combining site (antigenic determinant group); in addition, the corresponding antiserum may contain many types of antibodies differing in their specificities.

While a full description and discussion of serologic techniques is beyond the scope of this laboratory, the three simple procedures selected for this experiment can be performed without elaborate equipment or advanced training. A full understanding of how these techniques work, however, will undoubtedly require additional reading either in your microbiology text or some standard immunology text.

PART A. THE SLIDE AGGLUTINATION REACTION

The antigen-antiserum system used for this portion of the work will depend, in part, on what materials are readily available to your instructor. Antisera may have to be purchased or obtained through the immunization of rabbits or other animals (can be class project). Suggested alternatives would be the purchase of commercial blood-typing antisera (and let the students type their own blood) or the purchase of antisera (from Difco, BioQuest or others) specific for the strain of *Salmonella* used in the last experiment. The laboratory may be set up as a demonstration or as individual experiments with unknowns.

Some kinds of pathogenic bacteria containing only one major surface antigen will elicit antibody that can be tested for by the use of a killed bacterial suspension (whole-cell antigen) in a simple agglutination test. This is true of *Brucella abortus,* which can cause brucellosis, often called Bang's disease in cattle and undulant fever in man. Thus, a killed suspension of *Brucella abortus* can be used to test for the presence of agglutinins (agglutinating antibodies) specific for *Brucella* in the serum of humans or cattle or in the milk of cows. Presence of agglutinins would indicate that the individual is currently infected with *Brucella* or has recently recovered from an infection. Alternatively, antisera known to contain antibodies to *B. abortus* can be used as an aid in the identification of an unknown culture suspected to be *Brucella.* (The problem of serologic cross reactions between species of *Brucella* need not be dealt with here.)

MATERIALS

1. Antiserum in dropper bottles. Obtained from infected cattle and diluted by instructor with normal saline (0.85% NaCl) to conserve antiserum, but will still give positive test in 10 min.

2. Unknown antiserum in dropper bottles. Prepared as above.

3. Control serum in dropper bottles. Normal cow serum diluted as above.

4. Antigen. Preferably standardized Brucella Antigen for Ring Test. This is a dilute, killed suspension of *B. abortus* containing a dye (methylene blue) to make the reaction more visible. Can often be obtained through regional offices of USDA Agricultural Research Service, or from the National Animal Disease Laboratory, Ames, Iowa, 50010.

5. Glass slides (1″ × 3″).

6. Toothpicks.

PROCEDURE

1. Select a clean slide and place one drop of control serum on one end. Place one drop of antiserum at the opposite end of the slide. Do not allow them to touch each other. Your partner will work with control serum and unknown serum (or whey from milk).

2. Add one drop of *Brucella* antigen to each drop of control serum. Do not allow dropper to touch serum.

3. Mix control serum and antigen together with a toothpick to get a uniform suspension. Keep the liquid confined to a small area or the suspension will dry out. Repeat with antiserum or unknown.

4. Periodically, tilt the slide back and forth to promote gentle mixing and to keep the cells in suspension.

5. Observe slide for signs of clumping over a 10-minute period. Clumping is detected by the granular appearance of the suspension in contrast to the even dispersion of cells in the control serum. Record results in appropriate section.

6. Time and materials permitting, your instructor will demonstrate:

   a. The ring test as used on pooled whole milk samples to detect infected herds (will require Brucella Ring Test Antigen and whole milk from infected cows).

   b. The tube agglutination test used to determine the agglutinin titer of serum (will require *Brucella* Tube Test Antigen and sera from infected cows, plus tubes, rack, etc.).

   c. The "rapid plate" method for estimating the agglutinin titer of serum (will require Brucella Rapid Plate Antigen and sera from infected cows, plus appropriate viewing box, etc).

PART B. CAPILLARY RING TEST FOR PRECIPITATING ANTIBODY

Precipitation, rather than agglutination, may be observed whenever antibody is mixed with soluble, instead of particulate, antigen, (for example, bacterial protein, albumin, or serum from other species.) Depending on the procedure chosen, the precipitate may form in agar (see next section), in a test tube, or at a liquid interface in a capillary tube. The procedure may be either quantitative or qualitative.

As a qualitative test, the capillary ring test has several advantages; it requires only small amounts of reactants (antigens and antibodies) and simple equipment, and the results are obtained in only a few minutes. Suppose, for example, that you have been given a sample of hamburger suspected of being adulterated with horsemeat. You are asked to substantiate or refute the allegation by providing scientific evidence.

MATERIALS

1. Control antigen (cow). Supernate from saline extract of pure ground beef. Preparation not critical, but solution should not be too turbid. (To obtain best results adjust concentration with normal saline in trials made before class.)

2. Control antigen (horse). Supernate from saline extract of ground horse meat, prepared as above. If necessary, substitute diluted horse serum obtained from a commercial source.

3. Unknown antigen. Supernate from saline extract of ground beef suspected of being adultered (in this case, by your instructor). Prepare as above.

4. Anti-horse antiserum. Purchased or prepared by immunizing a rabbit or goat with horse serum. Any cross-reacting antibodies to cow to be removed by adsorption (see any standard text). Diluted one half with normal saline.

NOTE: If desired, another antiserum-soluble antigen system may be substituted for the one described here; the principle remains the same.

5. Four capillary tubes (per pair of students), measuring 4 cm × 3mm (I.D.). Small block of wood drilled to hold tubes.

6. Disposable capillary (blood transfer) pipettes.

PROCEDURE

1. Using the capillary pipette, transfer enough diluted antiserum to each of two capillary tubes to fill one third full. There must be no air bubbles in the capillary tubes. If necessary, suck out the serum and start over.

2. Carefully layer the control antigen (cow) on top of the serum in *one* of the tubes as follows: (The objective is to create a sharp boundary or interface between the two reactants. The difference in the density of the two liquids helps to achieve this.)

   a. Using a clean capillary pipette, suck up a small amount of the cow antigen (no air bubbles).

b. Hold the tip of the pipette against the inside wall of the capillary tube, just above the serum meniscus.
c. Let the antigen solution flow out slowly to prevent mixing. Bring level up to two thirds full mark.

3. Layer the second capillary tube with the control antigen (horse) in the same manner. Your partner will prepare one tube with cow control antigen and another with the unknown.

4. Wipe fingerprints off the capillary tubes and observe the interface periodically over the next 10 minutes (preferably against a dark background with light shining on the side of the tube).

5. A white ring of precipitate developing at the interface indicates that the antigen and antibody are specific for each other. Compare results with your partner and record them in the proper section.

PART C. IMMUNODIFFUSION

Immunodiffusion involves precipitation in a semisolid medium such as agar gel, starch gel, or cellulose acetate. The object of the procedure is to bring together, through diffusion, optimal concentrations of antigen and antibody to form visible lines or bands of precipitation. This method can provide excellent resolution of the individual components of the reactants and is, therefore, a valuable guide to the identification of either antigens or antibodies. The number of precipitin lines or bands formed indicates the *minimal* number of antigen-antibody systems present, provided that artifacts due to temperature and to excesses of either reactant are excluded. The site of a precipitin line is influenced by the reactants' diffusibilities and their initial concentrations.

There are three widely used immunodiffusion methods (Fig. 34-1). The Oudin tube method involves the one-dimensional diffusion of one reactant. Agar containing antiserum is placed in the bottom of a small tube and fluid antigen is layered on top. As the antigen diffuses into the agar, a band(s) of precipitate forms. Each of the one or more bands which develop normally represents a single antigen precipitated by its specific antibody. The Preer method is also one-dimensional, but both antigen and antibody are free to diffuse. Undiluted antiserum is placed in the bottom of a tube; a small amount of

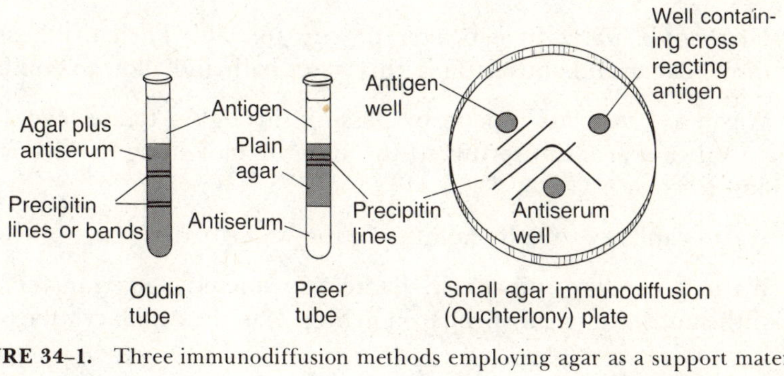

FIGURE 34–1. Three immunodiffusion methods employing agar as a support material.

agar is layered over the antiserum, and a solution of antigen is placed over the agar. This method involves double diffusion since the precipitin lines or bands develop in the central agar layer where neither of the reactants were originally present. The third method is the agar plate technique of Ouchterlony. It is essentially a two-dimensional double diffusion method for comparing liquid antigen and antibody preparations. After sufficient time passes for the reactants to diffuse from precut agar wells and form precipitin lines, a qualitative comparison of the type(s) of interaction(s) is made. The possible complexities introduced both by antigen-antibody interactions and by environmental factors (i.e., temperature variations, ionic strength, pH, etc.), preclude a casual or simplistic approach to the interpretation of results. Nonetheless, the Ouchterlony method provides excellent resolution as well as some information not obtainable by the other two methods. See text for further details. The Oudin procedure was chosen for this experiment because of its simplicity.

MATERIALS

1. Anti-horse serum (from Part B).

2. Control antigen (horse). Also from Part B but to which 0.01% merthiolate powder has been added as a preservative.

3. Washed agar (as in Experiment 13). Provide 1 tube containing 1 ml of molten agar (2.0% washed agar and 0.2% NaCl in distilled water).

4. Disposable capillary (blood transfer) pipettes.

5. Capillary tube (as in Part B) and wood holder.

6. 1 ml glass pipette.

7. Small piece of aluminum foil.

PROCEDURE

1. The only difficulty with this procedure is to prevent the molten agar from solidifying before you are ready. Thus, all glassware and reagents that will contact the molten agar should first be heated to 50 to 55C.

2. Heat some water in a beaker or cup to 55C. Turn off heat. Place tubes of molten agar and antiserum in this water bath and allow to equilibrate.

3. Warm a 1 ml glass pipette by passing through a flame several times. Transfer 1 ml of warm antiserum to the tube of molten agar. Mix well, but do not allow to cool.

4. Warm capillary tube by holding in the water bath.

5. Warm a capillary pipette in flame and immediately transfer a small amount of the antiserum agar (prepared in Step 3) to the capillary tube (enough

to fill two thirds). Avoid air bubbles or touching the wall above what will become the agar boundary. If necessary, start over with a clean capillary tube.

6. With a clean pipette, add sufficient antigen to fill the capillary tube (as shown in Fig. 34-1). Cover the top of the tube with a small piece of foil.

7. Store the capillary tube in your desk and record your observations every 2 days for 1 week.

RESULTS

PART A. SLIDE AGGLUTINATION REACTION

1. List whether agglutination occurred, and if so, the time required.
   a. Antiserum from infected cattle ______________________
   b. Control serum ______________________
   c. Unknown serum ______________________

PART B. CAPILLARY RING TEST

1. List whether precipitation occurred, and if so, the time required.
   a. Control antigen (cow) ______________________
   b. Control antigen (horse) ______________________
   c. Unknown antigen ______________________

PART C. IMMUNODIFFUSION

1. Record your results by making a simple drawing for each observation period, showing the number of precipitin lines and their location.

DISCUSSION

You are encouraged to use your text in answering the following questions.

1. Some strains of bacteria agglutinate spontaneously in saline and thus are not suitable for slide agglutination reactions. How can you determine whether you have been given such a strain? Is the control serum *always* necessary?

2. What is the basis of blood typing? What is the Rh factor?

3. What advantage would the tube agglutination test have over the slide agglutination test?

4. Some types of bacteria have capsules or slime layers which block agglutination reactions. Can you cite an example?

5. How does the milk ring test for brucellosis differ from the capillary ring test? Are they related in any way?

6. Why was the control antigen (horse) necessary for the capillary ring test?

7. Did you obtain evidence of adulteration of the hamburger? Does adulteration present a health problem? Why?

8. What advantages and disadvantages does the Oudin tube have when compared with the capillary ring test?

SECTION

VII

BACTERIAL TAXONOMY

EXPERIMENT 35

Identification and Classification (Taxonomy) of Unknown Bacteria

The first step in an exact study of any unknown bacterium is to determine its purity. The method by which you obtain only one kind of bacterium is known as isolation of organisms in pure culture. The pure culture so obtained is maintained in pure form throughout the study by rigorous application of aseptic technique.

The two most frequently used references on systematic identification and classification for beginning bacteriologists are *Bergey's Manual of Determinative Bacteriology* and *A Guide to the Identification of the Genera of Bacteria,* 2nd ed., by V. B. D. Skerman. *Bergey's Manual* is based on the *International Code of Nomenclature of the Bacteria and Viruses.*

You have obtained much information regarding the morphologic, cultural, and physiologic characteristics of your unknown since the beginning of this course. Such studies are frequently referred to as "systematic study of unknown microorganisms." By utilizing the information gathered by systematic study, you may now classify your unknown bacterium.

If you are using the seventh edition of *Bergey's Manual,* your isolate may belong to one of ten orders outlined in Class Schizomycetes. See Table on following page.

Your task is to select the one order which most resembles your unknown isolate. This task may appear not only monumental but also impossible; however, with proper use and understanding of the principles of taxonomy you may be able to identify your unknown isolate with relative ease.

Identification of an unknown isolate by use of the seventh edition of *Bergey's Manual* may be approached in two ways. First, carefully read the descriptions of the ten orders and select the one that agrees most with your unknown. After you have selected the order, you will be referred to a key of the families of that order. Again, you will select the one family that your unknown most resembles. Repeat the procedure and continue to narrow the choice by moving from family to tribe to genus and to species. You will find a vast amount of information regarding the morphologic, cultural, and physiologic characteristics of the *type species* within the genus. Select the one species that resembles most closely your unknown.

Keys to the Orders of Class Schizomycetes*

I. Cells rigid. Spherical, rod-shaped (straight or curved), or spiral in form. Sometimes in trichomes. Motile by means of polar flagella or nonmotile.
  A. Cells coccoid, straight or curved rods, or spiral in form. Sometimes occur as chains of cells. Cells may contain photosynthetic purple or green pigments. Not in trichomes. Usually motile by means of polar flagella. Occasionally nonmotile.
  Order I. Pseudomonadales, p. 35.
  B. Not as above.
    1. Cells in trichomes that are frequently in a sheath. Occasionally motile (swarm spores) or nonmotile conidia are developed. The sheaths may contain a deposit of ferric hydroxide, and the trichomes may be attached to a substrate.
    Order II. Chlamydobacteriales, p. 262.
    2. Cells reproduce by a process of budding rather than by ordinary cell divisions (fission). May be attached to a substrate by a stalk. One genus contains species with photosynthetic pigments *(Rhodomicrobium).*
    Order III. Hyphomicrobiales, p. 276.

II. Not as above.
  A. Cells rigid. Sperical or straight rod-shaped cells. Occur singly, in chains, or in trichomes. Motile by means of peritrichous flagella or nonmotile. Not acid-fast.
    1. Cells spherical or rod-shaped; no trichomes, though chains of cells may occur.
    Order IV. Eubacteriales, p. 281.
    2. Cells in trichomes.
    Order VI. Caryophanales, p. 830.
  B. Not as above.
    1. Cells rigid and may grow out into a branching mycelium-like structure which may even develop chains of aerial conidia, giving colonies a superficial resemblance to mold colonies. In two genera spores develop within sporangia (sporangiospores), and in one of these genera the spores are motile. Where cells occur singly or in simple branched forms, they are frequently acid-fast.
    Order V. Actinomycetales, p. 694.
    2. Not as above.
      a. Cells rigid, usually large, and may occur as coccoid cells or trichomes. Sulfur granules may occur on the surface or within the cells. Move by a gliding, oscillating or rolling, jerky motion like that of some blue-green algae. No flagella present.
      Order VII. Beggiatoales, p. 837.
      aa. Not as above.
        b. Longer or shorter flexuous cells.
          c. Cells flexuous, creeping on a substrate. Frequently pointed at both ends. Fruiting bodies are usually developed from a thin spreading colony (pseudoplasmodium). Slime bacteria.
          Order VIII. Myxobacterales, p. 854.
          cc. Cells in the form of longer or shorter spirals. Swim freely by flexion of cells.
          Order IX. Spirochaetales, p. 892.
        bb. Nonmotile, highly pleomorphic organisms of a very delicate character. Possess filterable stages.
        Order X. Mycoplasmatales, p. 914.

(*From *Bergey's Manual of Determinative Bacteriology,* 7th ed. Baltimore: Williams & Wilkins Co., 1957.)

Appendix A

PURE CULTURES OF MICROORGANISMS USED IN THIS COURSE

Many of the cultures can be purchased from the American Type Culture Collection (ATCC) or Northern Regional Research Laboratory (NRRL), or any other culture collection center.

Name of Organism	Experiment Number
Agrobacterium tumefaciens	31
Alcaligenes viscolactis	15, 30
Aspergillus niger	7, 29
Azotobacter vinelandii	26
Bacillus cereus	1, 4, 5, 6, 14, 18 19, 20, 24, 30, 32
Bacillus megaterium	19
Bacillus polymyxa	5
Bacillus subtilis	16
Brucella abortus (antigen)	33
Clostridium sporogenes	16, 24
Enterobacter (Aerobacter) aerogenes	5, 11, 14, 15, 30
Erwinia atroseptica	31
Erwinia carotovora	31
Escherichia coli	1, 4, 13, 17, 18 20, 25, 33
Escherichia coli B	12, 23
Escherichia coli K	12
Escherichia coli (thr^+ met^-)	22
Escherichia coli (thr^- met^+)	22
Geotrichum candidum	7
Lactobacillus arabinosus	15
Micrococcus albus	32
Micrococcus conglomeratus	32
Micrococcus flavus	21
Mycobacterium avium	4
Mycobacterium smegmatis	4
Mycobacterium tuberculosis BCG	4
Neisseria catarrhalis	13
Penicillium chrysogenum	7, 18
Penicillium notatum	18
Proteus vulgaris	6, 11, 25, 33
Pseudomonas aeruginosa	13, 18, 21, 26, 33

Name of Organism	Experiment Number
Pseudomonas fluorescens	11, 12, 14, 21, 24, 29
Rhizobium sp	26
Rhizopus nigricans	7
Saccharomyces cerevisiae	7, 15, 29
Salmonella enteritidis ser *typhimurium*	33
Sarcina lutea	19
Serratia marcescens	19, 21
Shigella flexneri	33
Spirillum serpens	1, 4, 5
Staphylococcus aureus	1, 4, 6, 11, 12, 13, 16, 18, 20, 24, 25, 32
Staphylococcus epidermidis	5
Streptococcus faecalis	25, 32
Streptococcus faecalis var. *liquefaciens*	30
Streptococcus lactis	13, 30
Streptococcus pyogenes	32
Streptomyces griseus	7, 18
Torulopsis lactis-condensi	30
Torulopsis sphericus	15
Torulopsis utilis	5
Bacteriophage T_3	12

Appendix B

STAINS AND REAGENTS

1. Stains and reagents for Gram stain

   a. Gram's crystal violet, Hucker's modification
   Solution A
   crystal violet (90% dye content) 2.0 g
   ethyl alcohol 20 ml
   Solution B
   ammonium oxalate 0.8 g
   distilled water 80 ml
   Mix solutions A and B.

   b. Gram's iodine (Gram's modification of Lugol's solution)
   Iodine 1.0 g
   KI 2.0 g
   Distilled water 300 ml
   Add iodine after KI is dissolved in water.

   c. Gram's alcohol (decolorizer)
   95% ethyl alcohol 98 ml
   Acetone 2 ml

   d. Gram's safranin (counterstain)
   Safranin O (2.5% solution in 95% ethyl alcohol) 10 ml
   Distilled water 100 ml

2. Löffler's alkaline methylene blue
   Solution A
   Methylene blue (90% dye content) 0.3 g
   Ethyl alcohol (95%) 30 ml
   Solution B
   Dilute KOH (0.01% by weight) 100 ml
   Mix solutions A and B.

3. Aqueous methylene blue (methylene blue in dilute alcohol)
   Methylene blue (80% dye content) 0.3 g
   Ethyl alcohol (95%) 30 ml
   Distilled water 100 ml
   Dissolve the dye in alcohol, then add water.

4. Ziehl's carbol-fuchsin (acid-fast staining)
   Solution A
   Basic fuchsin (90% dye content) 0.3 g
   Ethyl alcohol (95%) 10 ml
   Solution B
   Phenol 5 g
   Distilled water 95 ml
   Mix solutions A and B.

5. Negative stain (Dorner's nigrosin solution)

Nigrosin, water soluble	10 g
Distilled water	100 ml

Filter twice through double filter paper and the solution may be preserved by adding 0.5 ml formalin.

6. Malachite green (spore stain, Conklin's modification of Wirtz method)

Malachite green	5 g
Distilled water	95 ml

Filter before use.

7. Sudan black-B (staining for fat droplets, Burdon's method)

Sudan black-B	0.3 g
70% ethyl alcohol	100 ml

Mix thoroughly and allow to stand overnight.

8. Acid alcohol

Concentrated HCl	3 ml
95% ethyl alcohol	100 ml

9. Alcian blue capsule stain, 1% aqueous solution
10. Xylene or xylol
11. Phosphomolybdic acid solution (1%)
12. Methyl green, 1% aqueous solution
13. India ink
14. Vials of petroleum jelly
15. HCl, 1% aqueous solution
16. Acid $HgCl_2$

$HgCl_2$	15 g
Conc. HCl	20 ml

Mix, then add to 100 ml distilled water.

17. $CuSO_4$, saturated
18. H_2O_2, 3% aqueous solution
19. Methyl red indicator

Methyl red	0.1 g

Dissolve in 300 ml 95% ethyl alcohol.
Dilute to 500 ml with distilled water.

20. KOH, 40% aqueous solution
21. Phenol, various concentrations
22. Roccal (alkyldimethylbenzylamine chloride), Winthrop-Sterling Drug
23. Assorted commercial bactericides
24. Assorted commercially prepared antibiotic disks
25. 70% ethyl alcohol

26. Nessler's reagent for testing ammonia
    a. KI 50 g
       Distilled water (ammonia-free) 35 ml
    b. Add a saturated aqueous solution of $HgCl_2$ until a slight precipitate persists.
    c. Add 400 ml of 50% aqueous solution of KOH.
    d. Dilute to 1000 ml with ammonia-free distilled water.
    e. Allow to stand for a week, decant supernatant liquid, keep in dark-colored, tightly-capped bottle to avoid photooxidation.

27. Reagents for testing nitrites

    Reagent 1

    Sulfanilic acid 2 g

    5N acetic acid 250 ml (1 part glacial acetic acid to 2.5 parts water)

    Mix the two reagents thoroughly, keep in dark-colored bottle.

    Reagent 2

    Dimethyl-α-naphthylamine 1.5 ml

    5N acetic acid 250 ml

    Mix the two reagents thoroughly, keep in dark-colored bottle.

28. Antazoline reagent for testing nitrates

    2-(N-benzylanilinomethyl)iminazoline (antazoline) 1 g

    Distilled water 99 ml

    Keep in cold storage until used.

    (Chemical may be obtained from CIBA Pharmaceutical Co., Summit, N.J.)

29. NaOH, 0.1N

30. Methylene blue thiocyanate

    1:30,000 aqueous solution

31. Horse serum; anti-horse serum

32. Bovine serum; anti-bovine serum

Appendix C

MEDIA

Most media used in this manual are available in dehydrated form.

1. Ammonium medium (Exp. 26)

$(NH_4)_2SO_4$	2.0 g
K_2HPO_4	1.0 g
$MgSO_4$	0.5 g
$FeSO_4$	0.1 g
$MgCO_3$	5.0 g
Distilled water	1000 ml

2. Antibiotic assay agar (Exp. 18)

Tryptone	5.0 g
Peptone	5.0 g
Yeast extract	3.0 g
Beef extract	1.0 g
Glucose	2.0 g
Agar	15.0 g
Distilled water	1000 ml

3. Antibiotic assay soft agar (Exp. 18)
   The same as antibiotic assay agar above, except reduce agar to 7.0 g per liter.

4. Arginine broth (Exp. 25)

Peptone	1.0 g
NaCl	5.0 g
K_2HPO_4	0.3 g
Glucose	1.0 g
Arginine	10.0 g
Distilled water	1000 ml
Brom thymol blue	1 ml of 1.6% alcoholic solution

5. Blood agar (Exp. 32)

Blood agar base (commercially available)	40 g
Distilled water	1000 ml

   Sterilize, cool to 45C, add 5% fresh defibrinated blood, mix thoroughly, and pour into sterile Petri dishes.

6. Brain heart infusion agar (commercially available) (Exp. 13)

7. Butterfat agar (Exp. 14)

Beef extract	3.0 g
Peptone	5.0 g
Agar	15.0 g
Butterfat	10.0 g
Distilled water	1000 ml

Make certain that lipids are well homogenized with the remaining nutrients.

8. Casein agar (Exps. 14 and 24)

Sodium caseinate	2.0 g
Glucose	1.0 g
K_2HPO_4	0.2 g
$MgSO_4$	0.2 g
$FeSO_4$	trace
Agar	15.0 g
Distilled water	1000 ml

9. Casein broth (Exp. 26)
   Use same ingredients as for casein agar, except omit agar.

10. Endo agar (Exp. 28)
    Medium is commercially available in dehydrated form.

11. mEndo broth (Exp. 28)
    Medium is commercially available in dehydrated form.

12. mEnterococcus agar (Exp. 32)
    Medium is commercially available in dehydrated form.

13. Eosin methylene blue (EMB) agar (Exps. 28, 33)
    Medium is commercially available in dehydrated form.

14. mFC broth (Exp. 28)
    Medium is commercially available in dehydrated form.

15. Fermentation broths (Exp. 15)
    Tryptone-yeast extract medium (TYE)

Tryptone	10.0 g
Yeast extract	3.0 g
Brom cresol purple indicator (BCP)	0.04% (or add 1 ml of 1.6% alcoholic solution)
Distilled water	1000 ml

Glucose broth:	TYE medium + 10.0 g glucose (for Smith fermentation tubes)
Sucrose broth:	TYE medium + 5.0 g sucrose
Glucose broth:	TYE medium + 5.0 g glucose (for Durham tubes)
Xylose broth:	TYE medium + 5.0 g xylose
Lactose broth:	TYE medium + 5.0 g lactose

16. Fluid thioglycollate broth (Exps. 11, 16, 24)
    Medium is commercially available in dehydrated form.

17. Gelatin agar (Exp. 14)

Beef extract	3.0 g
Peptone	5.0 g
Gelatin	4.0 g
Agar	15.0 g
Distilled water	1000 ml

18. Glucose ammonium acid phosphate agar (Exp. 32)

Glucose	10.0 g
$NH_4H_2PO_4$	1.0 g

KCl	0.2 g
$MgSO_4$	0.2 g
Agar	15.0 g
Distilled water	1000 ml
Brom cresol purple indicator	1.0 ml of 1.6% alcoholic solution per liter

19. Glucose–peptone–yeast extract–soft agar (Exp. 11)

Glucose	2.0 g
Peptone	3.0 g
Yeast extract	2.0 g
Agar	5.0 g
Distilled water	1000 ml

20. Glucose–yeast extract–agar (Exp. 7)

Peptone	3.0 g
Yeast extract	2.0 g
KH_2PO_4	0.5 g
K_2HPO_4	1.0 g
Glucose	10.0 g
Agar	15.0 g
Distilled water	1000 ml

21. Glycerol yeast extract nutrient agar (Exp. 23)

Beef extract	3.0 g
Peptone	5.0 g
Glycerol	20 ml
Yeast extract	2.0 g
Agar	15.0 g
Distilled water	1000 ml

22. King's medium (Exp. 21)

Peptone	20.0 g
Glycerol	10.0 ml
K_2SO_4	10.0 g
$MgCl_2$	1.4 g
Agar	15.0 g
Distilled water	1000 ml

23. Lactose broth (Exps. 20, 25)

Beef extract	3.0 g
Peptone	5.0 g
Lactose	10.0 g
Distilled water	1000 ml
Brom thymol blue indicator	1.0 ml of 1.6% alcoholic solution

24. Lauryl tryptose broth (Exps. 27, 28)
    Medium is commercially available in dehydrated form.

25. Litmus milk (Exp. 30)

Dry milk	100 g
Litmus	0.75 g
Distilled water	1000 ml

Litmus should be ground with mortar and pestle.

26. Lysine-iron-agar (Exp. 33)
    Medium is commercially available in dehydrated form.

27. MacConkey agar (Exp. 33)
    Medium is commercially available in dehydrated form.

28. Mannitol salt agar (Exp. 32)
    Medium is commercially available in dehydrated form.

29. Medium for low pH, high sugar tolerance (Exp. 29)
    Nutrient broth (control)
    Nutrient broth + citric acid (pH 3.0)
    Nutrient broth + 20% glucose
    Nutrient broth + 20% glucose + citric acid (pH 3.0)

30. Medium for salt tolerance (Exp. 29)
    Nutrient broth (control)
    Nutrient broth + 3% NaCl
    Nutrient broth + 6.5% NaCl
    Nutrient broth + 10% NaCl

31. Minimal medium (Exp. 22)

$(NH_4)_2SO_4$	1.0 g
Glucose	2.0 g
K_2HPO_4	7.0 g
$MgSO_4$	0.5 g
Agar	15.0 g
Distilled water	1000 ml

32. Motility agar (Exp. 6)

Glucose	2.0 g
Peptone	3.0 g
Yeast extract	2.0 g
Agar	5.0 g
Distilled water	1000 ml

33. Nitrite medium (Exp. 26)

$NaNO_2$	1.0 g
K_2HPO_4	1.0 g
$MgSO_4$	0.5 g
$FeSO_4$	0.1 g
$NaHCO_3$	1.0 g
Distilled water	1000 ml

34. Nitrogen-free medium (Exp. 26)

Mannitol	15.0 g
K_2HPO_4	0.5 g
NaCl	0.2 g
$MgSO_4 \cdot 7H_2O$	0.2 g
$CaCO_3$	5.0 g
$FeCl_3$	trace
$MnSO_4$	trace
Distilled water	1000 ml

35. Nutrient agar (used in many experiments)

Beef extract	3.0 g

Peptone	5.0 g
Agar	15.0 g
Distilled water	1000 ml

36. Nutrient agar, soft (Exp. 20)
    Same as above, except agar, 7.0 g

37. Nutrient broth (used in many experiments)

Beef extract	3.0 g
Peptone	5.0 g
Distilled water	1000 ml

38. Nutrient gelatin (Exp. 11)
    Medium is commercially available in dehydrated form.

39. Peptone broth (Exp. 26)

Peptone	10.0 g
K_2HPO_4	0.5 g
$MgSO_4$	0.2 g
Distilled water	1000 ml

40. Potato medium (Exp. 11)
    Wash potatoes, use large cork borer, place in test tube with 1 ml distilled water, autoclave for 5 minutes.

41. Sabouraud's dextrose agar (Exp. 7)
    Medium is commercially available in dehydrated form.

42. SS agar (Exp. 33)
    Medium is commercially available in dehydrated form.

43. Starch agar (Exp. 14)

Beef extract	3.0 g
Peptone	5.0 g
"Soluble starch"	4.0 g
Agar	15.0 g
Distilled water	1000 ml

44. Triple sugar-iron-agar (Exp. 33)
    Medium is commercially available in dehydrated form.

45. Tryptic soy agar (used in many experiments)
    Add desired amount of glucose to dehydrated medium.

46. Tryptic soy broth (used in many experiments)
    Add desired amount of glucose to dehydrated medium.

47. Tryptone broth (Exp. 12)

Tryptone	10.0 g
NaCl	5.0 g
Distilled water	1000 ml

48. Tryptone glucose extract agar (Exp. 30)

Beef extract	3.0 g
Yeast extract	2.0 g
Tryptone	5.0 g
Glucose	1.0 g
Agar	15.0 g
Distilled water	1000 ml

49. Tryptone soft agar (Exp. 12)

Tryptone broth + 7.0 g agar per liter

50. Tryptone yeast extract glucose agar (Exps. 22, 30)

Tryptone	10.0 g
Yeast extract	3.0 g
Glucose	1.0 g
NaCl	5.0 g
Agar	15.0 g
Distilled water	1000 ml

51. Tryptone–yeast extract–nitrate broth (Exp. 26)

Tryptone	15.0 g
KNO_3	1.0 g
Yeast extract	3.0 g
Distilled water	1000 ml

52. Urea broth (Exp. 26)

Yeast extract	0.1 g
KH_2PO_4	9.1 g
Na_2HPO_4	9.5 g
Urea	20.0 g
Distilled water	1000 ml

Filter sterilize and distribute into tubes under aseptic conditions.

53. Urea broth (Exp. 33)

Medium is commercially available in dehydrated form.

54. Yeast lysate–mannitol agar (Exp. 26)

Mannitol	10.0 g
K_2HPO_4	0.5 g
$MgSO_4$	0.2 g
NaCl	0.1 g
Agar	15.0 g
Yeast water*	100 ml
Distilled water	900 ml

55. Washed agar minimal medium (Exps. 13, 33)

1.5% washed agar† in double distilled water

1.5% washed agar in salts solution‡

1.5% washed agar and 0.5% glucose in salts solution

1.5% washed agar, 0.5% glucose, and 0.5% peptone in salts solution

---

*Yeast water is prepared by adding 100 g pressed, starch-free yeast to 1 liter distilled water. Heat for 3 hr in flowing steam, place in cotton-plugged flask, or sterilize in autoclave 90 min at 15 lb pressure, and allow to stand. Siphon off supernatant liquid, adjust to pH 6.6 to 6.8. (2 g/l of yeast extract may replace yeast water; however, the results obtained with yeast water appear to be superior to those obtained with yeast extract.)

†Washed agar should be prepared prior to class. The procedure is to suspend either Noble Agar (Difco) or Ionagar (Oxoid) in an excess of cold double distilled water (large Erlenmeyer flask with magnetic stirring bar). The suspension is kept cold and the wash water changed 5 to 6 times over a period of 48 hr. The agar is then air dried under gauze or in a dessicator jar under vacuum at 37C until suitable for use. The above procedure is recommended for consistent results.

‡The salts solution is designed to provide several of the key elements necessary for growth and contains per 100 ml:

$NH_4H_2PO_4$	0.1 g	NaCl	0.5 g
K_2HPO_4	0.1 g	Distilled water	100 ml
$MgSO_4$	0.02 g	pH adjusted to 7.0 to 7.2	

Appendix D

COMMON MATERIALS AND EQUIPMENT USED IN THE COURSE

1. Microscopes with low, high, and oil immersion objectives; immersion oil, lens paper; stage and ocular micrometers.

2. Nichrome wire inoculating needles and loops.

3. Microscope slides, cover glasses, hanging drop slides, porcelain spot plates, slide agglutination glass plates (with grids), slide holder and forceps.

4. Cakes of Bon Ami (for cleaning slides).

5. Bunsen burner, ring stand, wire gauze.

6. Thermometers (−10 to 110C).

7. Standard culture test tubes, Durham tubes, Petri dishes, flasks, bottles, tumblers, etc.

8. Smith fermentation tubes, Frost gasometer.

9. Membrane filter units and membranes.

10. Petri dish turntable and bent glass spreading rods.

11. Anaerobe jar, electrical or gas pack.

12. Pipettes (1.0 ml, 5.0 ml, and 10.0 ml) with a canister and dispenser.

13. Burette (50.0 ml), clamp, burette stand.

14. Colony counter.

15. pH meter.

16. Balance, weighing paper (bibulous).

17. Water bath (range 20 to 80C).

18. Magnetic stirrer.

19. Chemostat (demonstration use).

20. Hypodermic syringes (1 ml or 5 ml).

21. Rotary shaker.

22. Steamer.

23. Incubators (range: room temperature to 65C).

24. Dry heat sterilizer; steam heat sterilizer (autoclave).

25. Spectrophotometer.
26. Germicidal ultraviolet light (260 nm).
27. Photoflood lamp.
28. Solder and soldering gun.
29. Ice pick.
30. Wooden tongue blades and wood applicators.

Appendix E

MOST PROBABLE NUMBER INDEX

The following table represents the MPN Index and 95% confidence limits for various combinations of positive and negative results when five 10-ml portions, five 1-ml portions, and five 0.1-ml portions are used.*

No. of Tubes Giving Positive Reaction out of			MPN Index per 100 ml	95% Confidence Limits	
5 of 10 ml Each	*5 of 1 ml Each*	*5 of 0.1 ml Each*		*Lower*	*Upper*
0	0	0	<2		
0	0	1	2	<0.5	7
0	1	0	2	<0.5	7
0	2	0	4	<0.5	11
1	0	0	2	<0.5	7
1	0	1	4	<0.5	11
1	1	0	4	<0.5	11
1	1	1	6	<0.5	15
1	2	0	6	<0.5	15
2	0	0	5	<0.5	13
2	0	1	7	1	17
2	1	0	7	1	17
2	1	1	9	2	21
2	2	0	9	2	21
2	3	0	12	3	28
3	0	0	8	1	19
3	0	1	11	2	25
3	1	0	11	2	25
3	1	1	14	4	34
3	2	0	14	4	34
3	2	1	17	5	46
3	3	0	17	5	46

*The quantities indicated at the heads of the columns relate more specifically to finished waters. The values may be used in computing the MPN in larger or smaller portion plantings in the following manner: if, instead of portions of 10, 1.0, and 0.1 ml, a combination of portions of 100, 10, and 1 ml is used, the MPN is recorded as 0.1 times the value given in the applicable table. If, on the other hand, a combination of corresponding portions at 1.0, 0.1, and 0.01 ml is planted, record 10 times the value shown in the table; if a combination of portions of 0.1, 0.01, and 0.001 ml is planted, record 100 times the value shown in the table; and so on for other combinations.

(After *Standard Methods for the Examination of Water and Wastewater,* 13th ed. APHA, 1971.)

Table *(continued)*

No. of Tubes Giving Positive Reaction out of			MPN Index per 100 ml	95% Confidence Limits	
5 of 10 ml Each	*5 of 1 ml Each*	*5 of 0.1 ml Each*		*Lower*	*Upper*
4	0	0	13	3	31
4	0	1	17	5	46
4	1	0	17	5	46
4	1	1	21	7	63
4	1	2	26	9	78
4	2	0	22	7	67
4	2	1	26	9	78
4	3	0	27	9	80
4	3	1	33	11	93
4	4	0	34	12	93
5	0	0	23	7	70
5	0	1	31	11	89
5	0	2	43	15	110
5	1	0	33	11	93
5	1	1	46	16	120
5	1	2	63	21	150
5	2	0	49	17	130
5	2	1	70	23	170
5	2	2	94	28	220
5	3	0	79	25	190
5	3	1	110	31	250
5	3	2	140	37	340
5	3	3	180	44	500
5	4	0	130	35	300
5	4	1	170	43	490
5	4	2	220	57	700
5	4	3	280	90	850
5	4	4	350	120	1000
5	5	0	240	68	750
5	5	1	350	120	1000
5	5	2	540	180	1400
5	5	3	920	300	3200
5	5	4	1600	640	5800
5	5	5	⩾2400		

Appendix F

REFERENCES AND RECOMMENDED READING

Aaronson, S.: Experimental Microbial Ecology. Academic Press, New York. 1970.

Alexander, M.: Microbial Ecology. John Wiley and Sons, Inc., New York. 1971.

Alexopoulos, C.J., And Bold, H. C.: Algae and Fungi. The Macmillan Co., New York. 1967.

Blair, J. E., Lennette, E. H., and Truant, J. O.: Manual of Clinical Microbiology. American Society for Microbiology, Bethesda, Maryland. 1970.

Bodily, H. L., Updyke, E. L., and Mason, J. O. (Eds.): Diagnostic Procedures for Bacterial, Mycotic and Parasitic Infections, 5th ed. American Public Health Association, New York. 1970.

Braun, W.: Bacterial Genetics, 2nd ed. W. B. Saunders Co., Philadelphia. 1965.

Brock, T. D.: Principles of Microbial Ecology. Prentice-Hall, Inc., Englewood Cliffs, New Jersey. 1966.

Buchanan, R. E., and Gibbons, N. E.: Bergey's Manual of Determinative Bacteriology, 8th ed. The Williams & Wilkins Co., Baltimore. 1974.

Campbell, D. H.: Methods in Immunology, 2nd ed. W. A. Benjamin, Inc., New York. 1970.

Collins, C. H., and Lyne, P. M.: Microbial Methods, 3rd ed. University Park Press, Baltimore. 1970.

Commission of Editors of Biochemical Journals: Enzyme nomenclature. Science. *150:*719, 1965.

Conant, N. F., Smith, D. T., Baker, R. D., and Callaway, J. L.: Manual of Clinical Mycology, 3rd ed. W. B. Saunders Co., Philadelphia. 1971.

Cowan, S. T., and Steel, K. J.: Manual for the Identification of Medical Bacteria. Cambridge University Press, London. 1970.

Davis, B. D., Dulbecco, R., Eisen, H. N., Ginsberg, H. S., and Wood, W. B.: Microbiology, 2nd ed. Harper and Row, Hagerstown, Maryland. 1973.

Difco Manual of Dehydrated Culture Media and Reagents, 9th ed. Difco Laboratories, Detroit, Michigan. 1953.

Difco Supplementary Literature. Difco Laboratories, Detroit, Michigan. 1968.

Ecology of Soil-Borne Plant Pathogens. Prelude to Biological Control. University of California Press, Berkeley. 1970.

Edwards, P. R., and Ewing, W. H.: Identification of Enterobacteriaceae, 3rd ed. Burgess Publishing Co., Minneapolis. 1972.

Frazier, W. C.: Food Microbiology, 2nd ed. McGraw-Hill Book Co., New York. 1967.

Frobisher, M., Hinsdill, R. D., Crabtree, K. T., and Goodheart, C. R.: Fundamentals of Microbiology, 9th ed. W. B. Saunders Co., Philadelphia. 1974.

Garrod, L. P., and O'Grady, F.: Antibiotics and Chemotherapy, 3rd ed. The Williams & Wilkins Co., Baltimore. 1971.

Gibbs, B. M., and Shapton, D. A.: Identification Methods for Microbiologists, Part B. Academic Press, New York. 1968.

Gibbs, B. M., and Skinner, F. A.: Identification Methods for Microbiologists, Part A. Academic Press, New York. 1966.

Gilles, R. R., and Dodds, T. C.: Bacteriology Illustrated, 3rd ed. The Williams & Wilkins Co., Baltimore. 1973.

Goodheart, C. R.: An Introduction to Virology. W. B. Saunders Co., Philadelphia. 1969.

Graber, C. D.: Rapid Diagnostic Methods in Medical Microbiology. The Williams & Wilkins Co., Baltimore. 1970.

Isolation Techniques for Use in Hospitals. U. S. Department of Health, Education and Welfare, Center for Disease Control, Atlanta, Ga. 1970.

Lillie, R. D., et al.: H. J. Conn's Biological Stains, 8th ed. The Williams & Wilkins Co., Baltimore. 1969.

Martin, L. C.: The Theory of the Microscope. Elsevier Publishing Co., Inc., New York. 1966.

Meynell, G. G., and Meynell, E.: Theory and Practice in Experimental Bacteriology. Cambridge University Press, London. 1965.

Norris, J. R., and Ribbons, D. W.: Methods in Microbiology, Vols. 1–3. Academic Press, New York. 1969–1970.

Public Health Service, U. S. Department of Health, Education, and Welfare: Algae in Water Supplies. U. S. Government Printing Office, Washington D. C. 1969.

Public Health Service, U. S. Department of Health, Education, and Welfare: Milk Ordinance and Code. Publication No. 229. U. S. Government Printing Office, Washington, D. C. 1967.

Recommended Methods for the Microbiological Examination of Foods, 2nd ed. American Public Health Association, New York. 1966.

Rohde, P. A.: BBL Manual of Products and Laboratory Procedures, 5th ed. BioQuest, Division of Becton, Dickinson and Co., Cockeysville, Maryland. 1968.

Shapton, D. A., and Board, R. G.: Isolation Methods for Anaerobes. Academic Press, New York. 1971.

Shapton, D. A., and Gould, G. W.: Isolation Methods for Microbiologists. Academic Press, New York. 1969.

Skerman, V. B. D.: A Guide to the Identification of Genera of Bacteria, 2nd ed. The Williams & Wilkins Co., Baltimore. 1967.

Smith, G. M.: The Fresh-Water Algae of The United States. McGraw-Hill Book Co., New York. 1950.

Standard Methods for the Examination of Dairy Products, 12th ed. American Public Health Association, New York. 1967.

Stanier, R. Y., Doudoroff, M., and Adelberg, E. A.: The Microbial World, 3rd ed. Prentice-Hall, Inc., Englewood Cliffs, New Jersey. 1970.

Taras, M. J., Greenberg, A. E., Hoak, R. D., and Rand, M. C.: Standard Methods for the Examination of Water and Wastewater. American Public Health Association, Washington, D. C. 1971.